T0260743

Integration of Renewable Energy Sources with Smart Grid

Scrivener Publishing
100 Cummings Center, Suite 541J
Beverly, MA 01915-6106

Next-Generation Computing and Communication Engineering

Series Editors: Dr. G. R. Kanagachidambaresan and Dr. Kolla Bhanu Prakash

Developments in artificial intelligence are made more challenging because the involvement of multi-domain technology creates new problems for researchers. Therefore, in order to help meet the challenge, this book series concentrates on next generation computing and communication methodologies involving smart and ambient environment design. It is an effective publishing platform for monographs, handbooks, and edited volumes on Industry 4.0, agriculture, smart city development, new computing and communication paradigms. Although the series mainly focuses on design, it also addresses analytics and investigation of industry-related real-time problems.

Publishers at Scrivener
Martin Scrivener (martin@scrivenerpublishing.com)
Phillip Carmical (pcarmical@scrivenerpublishing.com)

Integration of Renewable Energy Sources with Smart Grid

Edited by

M. Kathiresh
A. Mahaboob Subahani
and
G.R. Kanagachidambaresan

Scrivener
Publishing

WILEY

This edition first published 2021 by John Wiley & Sons, Inc., 111 River Street, Hoboken, NJ 07030, USA and Scrivener Publishing LLC, 100 Cummings Center, Suite 541J, Beverly, MA 01915, USA
© 2021 Scrivener Publishing LLC
For more information about Scrivener publications please visit www.scrivenerpublishing.com.

Wiley Global Headquarters
111 River Street, Hoboken, NJ 07030, USA

For details of our global editorial offices, customer services, and more information about Wiley products visit us at www.wiley.com.

Limit of Liability/Disclaimer of Warranty
While the publisher and authors have used their best efforts in preparing this work, they make no representations or warranties with respect to the accuracy or completeness of the contents of this work and specifically disclaim all warranties, including without limitation any implied warranties of merchantability or fitness for a particular purpose. No warranty may be created or extended by sales representatives, written sales materials, or promotional statements for this work. The fact that an organization, website, or product is referred to in this work as a citation and/or potential source of further information does not mean that the publisher and authors endorse the information or services the organization, website, or product may provide or recommendations it may make. This work is sold with the understanding that the publisher is not engaged in rendering professional services. The advice and strategies contained herein may not be suitable for your situation. You should consult with a specialist where appropriate. Neither the publisher nor authors shall be liable for any loss of profit or any other commercial damages, including but not limited to special, incidental, consequential, or other damages. Further, readers should be aware that websites listed in this work may have changed or disappeared between when this work was written and when it is read.

Library of Congress Cataloging-in-Publication Data

ISBN 978-1-119-75042-0

Cover image: Pixabay.Com
Cover design by Russell Richardson

Set in size of 11pt and Minion Pro by Manila Typesetting Company, Makati, Philippines

10 9 8 7 6 5 4 3 2 1

Contents

Preface

As the electricity demand increases day by day, the dependence on fossil fuels and harmful emissions that affect health and the environment also increase. So, the best practice to overcome these problems is obtaining energy from the natural resources such as wind, solar, and tide, which are renewable in nature. With the traditional electric setup, it is highly impossible to achieve reliable and efficient power delivery. Thus, a new advanced version or modernization is necessary on the traditional electric grid architecture to overcome such glitches, and the solution is smart grid. The smart grid technology gives a roadmap for efficient, reliable, and clean future electric power delivery and also provides an opportunity for the effective integration of renewable energy sources. The intermittent energy generation of the renewable sources can be overcome by integrating energy storage systems such as battery energy storage system into the power grid. This integration needs a successful coordination between renewable power generation units, energy storage systems, and the power grid. Indeed, this coordination requires robust and innovative controls. The microgrid provides better opportunities for the consumers also to become a producer with highly efficient computing algorithms and triggering techniques. Furthermore, to reduce the stress on power system due to the uncertainty and inherent intermittence of renewable power generation units, the smart grid structure should consist of the main grid and multiple embedded microgrids.

This book starts with an overview of renewable energy technologies, smart grid technologies, and energy storage systems and covers the details of renewable energy integration with smart grid and the corresponding controls. This book provides better views on the power scenario in developing countries. The requirement of the integration of smart grid along with the energy storage systems is deeply discussed to acknowledge the importance of sustainable development of a smart city. The methodologies are made quite possible with highly efficient power convertor topologies and intelligent control schemes. These control schemes are capable of providing better control with the help of machine intelligence techniques and artificial intelligence. The book also addresses modern power convertor topologies and the corresponding control schemes for renewable energy integration with smart grid. The design and analysis of power converters that are used for the grid integration of solar PV along with simulation and experimental results are illustrated. The protection aspects of the microgrid with power electronic configurations for wind energy systems are elucidated. The book also discusses the challenges and mitigation measure in renewable energy integration with smart grid. This book serves to be a better material for the engineers and researchers working on microgrid, smart grid, and computing algorithms for converter and inverter circuits.

Sincere thanks to authors and reviewers for their best contributions in this book project. We would like to also thank Scrivener Publishing and Wiley for providing us a platform to share our knowledge and view towards sustainable development and renewable energy.

Sincere thanks to our parents, students, contributors, editor, and the Great Almighty.

The Editors
July 2021

Renewable Energy Technologies

V. Chamundeswari[1]*, R. Niraimathi[2], M. Shanthi[3]
and A. Mahaboob Subahani[4]

*[1]Department of EEE, St. Joseph's College of Engineering, Chennai,
Tamilnadu, India*
[2]Department of EEE, Mohamed Sathak Engineering College, Kilakarai, TN, India
*[3]Department of ECE, University College of Engg. Ramanathapuram,
Ramanathapuram, TN, India*
[4]Department of EEE, PSG College of Technology, Coimbatore, TN, India

Abstract

Most of the people around the world rely on the conventional energy sources such as oil, natural gas, and coal for their energy needs. Because of the fast depletion of these energy sources, there is a current global need for clean and renewable energy sources (RESs). The RESs are derived from natural sources such as the sun, wind, rain, tides of ocean, biomass, and geothermal. These are also referred as endless energy since they are replenished constantly. They are also considered as the most suitable energy sources for the future to achieve sustainable development, because the energy produced from these renewable sources does not harm the environment. In addition, they produce less pollutant while the energy conversion process.

Keywords: Renewable energy sources, solar, wind, hydro, tidal, geothermal

1. Introduction

Today's world is completely dependent on energy. As the demand for energy increases day by day and the conventional energy sources are depleting, there is an immediate need for finding out alternative energy sources. Hence, the contribution of renewable energy sources (RESs) in energy generation over the conventional energy sources has been

Corresponding author: chamuvins@gmail.com

M. Kathiresh A. Mahaboob Subahani and G.R. Kanagachidambaresan (eds.) Integration of Renewable Energy Sources with Smart Grid, (1–18) © 2021 Scrivener Publishing LLC

increasing year by year as shown in Figure 1.1. It is because of the reason that, the RESs are readily available and they are also sustainable. The energy from these sources is converted into a usable form and utilized for domestic as well as industrial applications. The renewable energies such as solar, wind, biomass, geothermal, hydro, and ocean energy can be converted into more useful energy like electricity. They deliver power with minimal impact on the environment. These sources are typically more green/cleaner than conventional energies like oil or coal. Among all the RESs, solar and wind energy plays a significant role in electric power generation [1]. They can supply power to either gird or isolated AC or DC loads [2]. Hydro energy is the next most used source for electricity generation. Geothermal energy which is produced from the heat of the earth's crust can also be used for energy conversion. Here, the thermal energy from the inner surface of the earth is converted into electricity. Tidal energy is also effectively utilized nowadays as low tide and high tide plays a vital role in producing electrical energy. All these RESs are discussed in detail in the following sections.

1.1   Types of Renewable Energy

Renewable energy includes:

1. Solar energy
2. Wind energy
3. Fuel cell

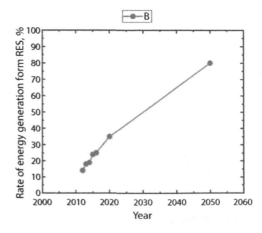

Figure 1.1 Increasing rate of energy generation from RES (present and future).

4. Biomass
5. Hydropower
6. Geothermal energy

The above-mentioned types of RESs are described along with their features as follows.

1.1.1   Solar Energy

The light energy produced from the sun is considered as one of the abundant and readily available energy resources. It is a significant source of renewable energy. The heat from the sun can also be extracted as thermal energy and used for solar-based heating applications. Depending on the type of energy capture and distribution, the solar power conversion technique is broadly classified as follows:

i.   Active solar technique
ii.   Passive solar technique

The active solar technique uses the concept of the solar photovoltaic (PV) system, concentrated solar power (CSP), and solar heating system, whereas the passive solar uses the technique of selecting materials of thermal nature and light dispersion property.

i. Active Solar Techniques

A. Solar Photovoltaic System
It works with the phenomenon of the PV effect, which is a combination of the physical and chemical process that generates voltage and current when light falls on a semiconducting device. In a semiconductor, conduction takes place when the electrons move from valence band to conduction band. There is some energy required for this operation. In a solar cell, the energy is produced from photons that are emitted from the sun. These photons help in moving the electrons from valence band to conduction band, thereby overcoming the gap between the bands. A photon incident on the surface could be reflected or transmitted. If it is reflected, then the electron cannot be dislodged. So, the photon must be absorbed to move the electrons from the valence band to the conduction band. Thus, the electron movement across the metallic junction takes place, creating a negative charge on one side with respect to the other. It is similar to a

battery with a negative terminal on one side and positive on the other side. The voltage and current are generated as long as light radiation is incident on the material. This effect only exists in the solar cells used in solar panels.

A solar panel is constructed by arranging PV cells or solar cells in series and parallel. A solar cell is a typical PN junction layer sandwiched as shown in Figure 1.2. Sunlight consists of photons or radiant solar energy. When the photons are incident on a PV cell, some get reflected and some get absorbed [3]. The absorbed photons aids in dislodging the electrons from the atoms of the solar cell material. These electrons move to the front surface of the solar cell and create an imbalance with respect to the back surface due to more flow of negative charge on one side. This imbalance results in a developed potential which creates electricity and a flow of current through an external load as shown in Figure 1.2. In general, a solar panel may have 60 cells connected together, yet, some solar panels are having even 72 cells also.

Monocrystalline and polycrystalline are the two major types of the solar cell. A monocrystalline solar cell is made from a single crystal of silicon, whereas polycrystalline cells are made by melting together many shards of silicon crystals. Monocrystalline solar cells are efficient when compared to polycrystalline type. It is due to the usage of monolithic crystal of silicon which aids in the easy flow of electrons that constitute the electric current. The electricity flow in polycrystalline silicon is very

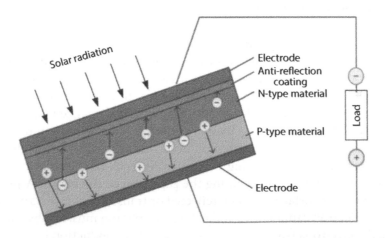

Figure 1.2 Structure of a solar cell.

difficult due to the many layers of silicon structure. The process of making solar panel using polycrystalline is very simple when compared to monocrystalline.

B. Concentrated Solar Power

The main objective of CSP is to focus the entire solar beam into a specific area. The heat energy thus produced in that area is converted into electricity. Other techniques developed based on the concept are parabolic trough system, dish system, and linear Fresnel collector. The concentrated solar beam produces heat energy which is used to drive the steam turbine and generate electricity.

Figure 1.3 shows the diagram of a CSP system. It uses lenses or mirrors to concentrate the major beam of light to concentrate on an area that is a receiver here. The light energy is converted into heat energy and it drives the steam turbine coupled with a generator and generates electricity. The following are the types of CPS system.

B1.1 Parabolic Trough Collector

Figure 1.4 shows the parabolic trough collector system which consists of a parabolic reflector that focuses the light onto a receiver aligned on the focal line of a reflector. The receiver is a tube which is filled with a working fluid and located above the reflector mirror arrangement. The working fluid is heated with the obtained light energy from the sun through the concentrator system [4].

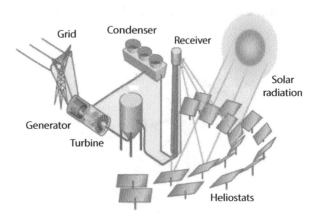

Figure 1.3 Concentrated solar power system.

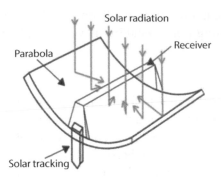

Figure 1.4 Parabolic trough collector.

B1.2 Parabolic Dish System

Figure 1.5 shows the parabolic dish system which consists of a parabolic dish concentrator to focus the solar beam. An axis tracking system to follow the sun's radiation is incorporated. The heat energy [5] from the concentrator is collected at the receiver side and used for generating electricity. The temperature at the dish can reach the maximum and can be used in solar reactors which are need for high temperature.

B1.3 Linear Fresnel System

Figure 1.6 shows the Fresnel reflector system. It uses flat mirrors to focus sunlight onto the receiver tubes which contains fluid in it. The diagram shows a primary and a secondary reflector system to make light energy completely fall on the receiving tubes [6]. As a result of it, the fluid is heated and steam produced drives the steam turbine. The generator coupled with the turbine generates electricity and fed to the loads. They are cheaper than

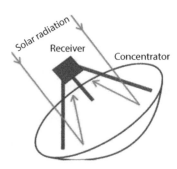

Figure 1.5 Parabolic dish system.

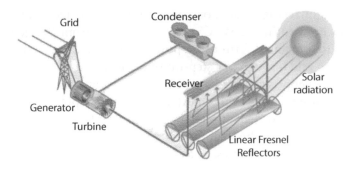

Figure 1.6 Linear Fresnel system.

the parabolic system and also captures more light energy from the sun. It can also be designed in various sizes.

Sometimes, the output yield is very low in this Fresnel system and so Fresnel reflectors with ray tracing was introduced to yield maximum output.

1.1.2 Wind Energy

In the current scenario, the wind energy system is one of the fastest-growing renewable energy. Wind turbine capacity has increased over time. In 1985, typical turbines had a rated capacity of 0.05 megawatts (MW) and a rotor diameter of 15 meters. Today's new wind power projects have turbine capacities of about 2-MW onshore and 3- to 5-MW offshore. Commercially available wind turbines have reached 10-MW capacity, with rotor diameters of up to 164 meters. The average capacity of wind turbines increased from 1.6 MW in 2009 to 2 MW in 2014 [7].

Wind energy conversion system (WECS) comprises of a wind turbine, gearbox, generator, converter, and transformers as shown in Figure 1.7. The wind energy or the kinetic energy is converted to mechanical energy using a wind turbine. The mechanical energy input is given to the generator and converted into electrical energy. Permanent magnet synchronous

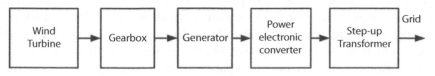

Figure 1.7 Wind energy conversion system (WECS).

generator, Squirrel cage induction generator or doubly fed induction generator can be used in the WECS. The AC output from the generator is converted to a required form using power electronic converter. It is then connected to grid through to a step-up transformer.

The wind energy is captured by the rotor blades and transferred to rotor hub. The rotating shaft provides mechanical energy input to the generator, which is further converted into electricity. The gear box helps in increasing the rotational speed of the shaft for the generator. The power extracted by the rotor blades may be expressed as follows:

$$P_T = 4\alpha(1-\alpha)^2 \frac{1}{2}\rho A u_0^3$$

where α is perturbation factor, ρ is density of the air, A is swept area of the blades, and u_o is speed of the upstream wind.

The wind turbines are largely classified into Horizontal Axis Wind Turbines (HAWT) and Vertical Axis Wind Turbines (VAWT). As the name implies, the HAWT has their blades rotating on an axis parallel to the ground. If the blades are placed in such a way that their rotational axis is perpendicular to the ground, it is called as VAWT. The HAWT can capable of producing more electricity as compared to VAWT. It is because the HAWT has more swept area than VAWT. Hence, the HAWT is generally preferred for commercial WECS. However, VAWT is used for small power applications.

1.1.3   Fuel Cell

In search of a clean energy source in the current energy sector, fuel cell has gained its importance. Fuel cell uses hydrogen as a fuel and the energy companies have started concentrating on low carbon hydrogen production. The industries have started using electrolyzer to produce clean hydrogen. In recent years, the electrolyzer installation has increased considerably. The survey says that 350,000 tonnes of low carbon hydrogen production has taken place by the end of year 2019 and 20 other new projects have been targeted by 2020. Fuel cell plays a vital role in generating electricity by using hydrogen as fuel. The more the hydrogen, the more the power. It is similar to a battery in some aspects but can supply energy for a long period of time and it is due to the continuous supply of fuel and oxygen to produce power. Due to these factors, fuel cell finds its application in satellites, manned spacecraft, and other relevant areas. It is also a type of RES that works on the principle of electrochemical reaction that converts chemical energy into an electrical energy. It converts the chemical energy of a fuel, namely, the hydrogen and an oxidizing agent, the oxygen, into electricity.

Figure 1.8 shows the diagram of a fuel cell. A fuel cell consists of an anode, cathode, and an electrolyte membrane. Hydrogen fuel is passed through the anode of a fuel cell and oxygen through the cathode. The hydrogen is split into electrons and protons at the anode side. The protons will pass through the membrane to the cathode side and the electrons are made to flow through an external circuit connected to the load After passing through the circuit, the electrons combine with the protons along with oxygen in air and produces water and heat as their by-product. Fuel cells are very clean as they use pure hydrogen as fuel. The efficiency of the fuel cell is high when compared to conventional techniques like steam turbine and internal combustion engine. The efficiency of a fuel cell can further be increased by interfacing it with a combined heat power system. The waste heat generated from the fuel cell can be used for various applications.

The types of fuel cell are as follows:

1. Proton exchange membrane (PEM) fuel cell
2. Direct methanol fuel cell (DMFC)
3. Alkaline fuel cell (AFC)
4. Phosphoric acid fuel cell (PAFC)
5. Molten carbonate fuel cell (MCFC)
6. Solid oxide fuel cell (SOFC)
7. Reversible fuel cell

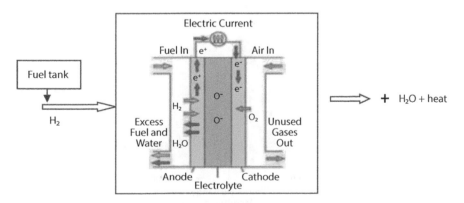

Figure 1.8 Diagram of a fuel cell.

1.1.3.1   Proton Exchange Membrane Fuel Cell

The frequently used fuel cell is PEM fuel cell. Figure 1.9 shows the PEM fuel cell. It is a light weight fuel cell and delivers high power density. It is also called as polymer electrolyte membrane (PEM) fuel cell [8]. It consists of carbon porous electrodes with solid polymer as an electrolyte and platinum as a catalyst. It operates with hydrogen, oxygen and water. Hydrogen fuel is given as an input from storage tanks. It operates at low temperatures and so considered as a durable one. A good catalyst is used but platinum is not so economical and it is sensitive to carbon monoxide poisoning. It requires a reactor to eradicate this poisoning effect and hence the cost also increases. Since it operates at low temperatures, its start-up time is very quick, and hence, it is suitable for automotive applications.

1.1.3.2   Direct Methanol Fuel Cell

Most of the fuel cells use hydrogen as the fuel to generate electricity, However, DMFC use methanol as a fuel input along with water. Methanol has higher energy density than hydrogen and it is easy to transport as it is like a liquid and similar to gasoline. It is a dense liquid but considered as a stable one. Its efficiency is around 40% and the operating temperature is between 50°C and 120°C. It is used as a powering circuit for laptops, cell phones, and other portable items.

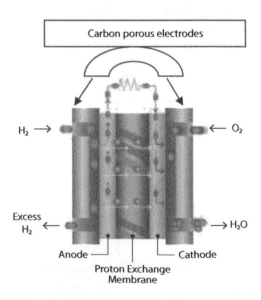

Figure 1.9 Proton exchange membrane (PEM) fuel cell.

1.1.3.3   Alkaline Fuel Cell

AFCs were the widely used fuel cells in space industry. Alkaline fuel cell is also called as "Bacon fuel cell" as it was invented by Francis Thomas Bacon. It is one of the considered fuel cell design. It is similar to PEM fuel cell except for the use of alkaline membrane instead of acid membrane. It uses a solution of potassium hydroxide in water as the electrolyte and non-precious metal as a catalyst at the anode and cathode. It uses hydrogen as fuel and pure oxygen to produce water and electricity. Because of its efficiency greater than 60%, it is used in space industries.

1.1.3.4   Phosphoric Acid Fuel Cell

It was the first commercial fuel cell in the mid-1960s. PAFC uses phosphoric acid as an electrolyte. The electrolyte is a pure or concentrated liquid phosphoric acid (H_3PO_4) in a silicon carbide matrix. It operates in the temperature range between 150°C and 210°C. Electrodes are made of carbon paper coated with platinum catalyst. It is used in buses and in stationary power generators in the range of 100 to 400 kW.

1.1.3.5   Molten Carbonate Fuel Cell

The conventional source–based power plants use MCFC for industrial and military applications. The electrolyte used in MCFC is a molten carbonate salt mixture immersed in ceramic matrix of beta alumina solid electrolyte. Because of its high operating temperature, metals are used as catalyst at the anode and cathode. It offers better efficiency when compared to PAFC which is around 65%. PAFC's efficiency is only 30% to 40%.

Solid oxide fuel cell (SOFC) and reversible fuel cells are the other types of fuel cell that are generally employed for various applications.

1.1.4   Biomass Energy

The energy derived from the organic matter of the living organism is the biomass. It is a RES that produces electricity with minimum cost. The organic material produced from plants and animals, crops and algae are used in biomass energy production. The global cumulative biomass energy generation is shown in Figure 1.10. When biomass is burned, heat is generated and the thermal energy is converted into electrical energy. This biomass can either be burned directly or converted into liquid biofuels or

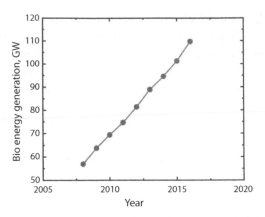

Figure 1.10 Cumulative bioenergy generation.

biogas. The conversion methods for biomass energy production include chemical, thermal, and bio-chemical [9].

Initially, direct combustion method was employed with wood as a fuel to produce energy. In the recent times, chemical treatments such as pyrolysis, fermentation, and anaerobic processes are implemented to convert these sources into a usable form such as ethanol. During pyrolysis treatment, coal is obtained as a product that strengthens the matter by burning it in the absence of oxygen.

The sources of biomass energy generation include the following:

i.   Wood and its processing waste: Heat energy is generated from the combustion of wood waste.
ii. Agricultural waste: It is burned as a fuel and it can be converted into liquid bio-fuels.
iii. Food and garbage waste: It is converted to bio-gas by landfill method or burned to generate electricity in power plants.
iv. Animal manure and sewage waste: it is converted to bio-gas.

1.1.4.1   Energy Production From Biomass

Solid biomass, such as wood and garbage, can be burned directly to produce heat. Biomass can also be converted into a gas called biogas or into liquid biofuels such as ethanol and biodiesel. These fuels can then be burned for energy production.

Biogas is formed when paper, food scraps, and yard waste are decomposed in landfills; it can also be produced by processing sewage and animal manure in special vessels called digesters. Ethanol is extracted from crops such as corn and sugar-cane by fermentation process. Biodiesel is produced from vegetable oils and animal fats and can be used in vehicles and as heating oil.

The current availability of biomass in India is estimated at about 500 million metric tonnes per year [10]. Studies from the Ministry has estimated surplus biomass availability of about 120–150 million metric tonnes per annum covering agricultural and forestry residues corresponding to a potential of about 18,000 MW. Apart from this, it is predicted that about 7,000-MW additional power could be generated through co-generation process.

1.1.5 Hydro-Electric Energy

As water is a never depleted source and the pressure of water is used to generate energy, hydro-electric power plants gained its significance in renewable energy industry. The energy of flowing water is converted into mechanical energy using a turbine and the coupled generator produced electricity from the mechanical energy.

Figure 1.11 shows hydro-electric power plant. The generator generates electricity by converting the input mechanical energy produced from the energy of water flow. Whenever a magnet moves past a conductor, it causes electricity and the flow of current exists. In a large generator, the electromagnets are made by circulating direct current through wire loops which

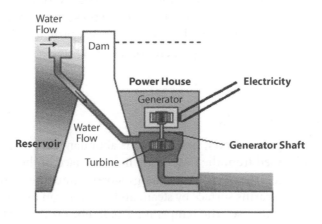

Figure 1.11 Typical hydro-electric power plant.

is wound on steel laminations. These are called as poles, which are held on the outer surface of the rotor. The rotor rotates at a fixed speed as it is connected to the turbine shaft. As the rotor rotates, it cuts the flux produced and induces an emf, and thus, a potential is developed across the generator output.

The reservoir system acts as a storage pump and can be used whenever required to generate electricity based on the demand. The construction of it is also very simple and this is one of the advantageous features of hydro-electric power generation [11].

1.1.5.1   Hydro Scenario

India is blessed with immense amount of hydro-electric potential and ranks fifth in terms of exploitable hydro-potential on global scenario. As per assessment made by CEA, India is endowed with economically exploitable hydro-power potential to the tune of 148,700 MW of installed capacity.

In 1998, Government of India announced "Policy on Hydro Power Development" under which impetus is given to development of hydropower in the country. This was a welcome step toward effective utilization of our water resources in the direction of hydropower development. During October 2001, Central Electricity Authority (CEA) came out with a ranking study which prioritized and ranked the future executable projects. As per the study, 399 hydro schemes with an aggregate installed capacity of 106,910 MW were ranked in A, B, and C categories depending upon their *inter se* attractiveness. During May 2003, Government of India launched 50,000-MW hydro initiative in which preparation of Pre-Feasibility Reports of 162 Projects totaling to 50,000 MW was taken up by CEA through various agencies. The PFRs for all these projects have already been prepared and projects with low tariff (first year tariff less than Rs.2.50/kWh) have been identified for preparation of DPR.

1.1.6   Geothermal Energy

As India is a tropical country, heat energy is abundant on our earth's crust. The energy obtained from the heat of the inner surface of the earth is geothermal energy. It is the unused heat energy stored under the earth's surface. It is carried to the earth's surface by steam and water. It can be used for heating and cooling purpose. The temperature gradient on the earth's surface with respect to the inner area is only used to generate electricity [12].

There are three types of geothermal power plants: dry steam, flash steam, and binary cycle.

(i) Dry Steam: Dry steam power plants draw from underground resources of steam. The steam is piped directly from underground wells to the power plant where it is directed into a turbine/generator unit.

(ii) Flash Steam: Flash steam power plants are the most common and use geothermal reservoirs of water with temperatures greater than 360°F (182°C). This very hot water flows up through wells in the ground under its own pressure. As it flows upward, the pressure decreases and some of the hot water boils into steam. The steam is then separated from the water and used to power a turbine/generator. Any leftover water and condensed steam are injected back into the reservoir, making this a sustainable resource.

(iii) Binary Steam: Binary cycle power plants operate on water at lower temperatures of about 225°F–360°F (107°C–182°C). Binary cycle plants use the heat from the hot water to boil a working fluid, usually an organic compound with a low boiling point. The working fluid is vaporized in a heat exchanger and used to turn a turbine. The water is then injected back into the ground to be reheated. The water and the working fluid are kept separated during the whole process, so there are little or no air emissions.

Currently, two types of geothermal resources can be used in binary cycle power plants to generate electricity: enhanced geothermal systems (EGSs) and low-temperature or co-produced resources [13].

a) Enhanced Geothermal Systems:
EGS provide geothermal power by tapping into the Earth's deep geothermal resources that are otherwise not economical due to lack of water, location, or rock type

b) Low-Temperature and Co-Produced Resources:
Low-temperature and co-produced geothermal resources are typically found at temperatures of 300°F (150°C) or less. Some low-temperature resources can be harnessed to generate electricity using binary cycle technology. Co-produced hot water is a by-product of oil and gas wells in the United States. This hot

water is being examined for its potential to produce electricity, helping to lower greenhouse gas emissions and extend the life of oil and gas fields.

1.1.6.1   Geothermal Provinces of India

In India, nearly 400 thermal springs occur (Satellites like the IRS-1 have played an important role, through infrared photographs of the ground, in locating geothermal areas. The Puga valley in the Ladakh region has the most promising geothermal field.), distributed in seven geothermal provinces. These provinces include The Himalayas: Sohana: West Coast; Cambay: Son-Narmada-Tapi (SONATA): Godavari and Mahanadi [14].

These springs are perennial and their surface temperature range from 37°C to 90°C with a cumulative surface discharge of over 1,000 L/m. The provinces are associated with major rifts or subduction tectonics and registered high heat flow and high geothermal gradient. For example, the heat flow values and thermal gradients of these provinces are 468 mW/m^2; 234°C/km (Himalayas); 93 mW/m^2; 70°C/km (Cambay); 120–260 mW/m^2; 60–90°C/km (SONATA); 129 mW/m^2; 59°C/km (West Coast); 104 mW/m^2; 60°C/km (Godavari) and 200 mW/m^2; 90°C/km (Bakreswar, Bihar).

The reservoir temperature estimated using the above described geothermometers are 120°C (West Coast), 150°C (Tattapani), and 200°C (Cambay). The depth of the reservoir in these provinces is at a depth of about 1 to 2 km.

These geothermal systems are liquid dominated and steam dominated systems prevail only in Himalayan and Tattapani geothermal provinces. The issuing temperature of water at Tattapani is 90°C, at Puga (Himalaya) is 98°C, and at Tuwa (Gujarat) is 98°C. The power-generating capacity of these thermal springs is about 10,000 MW (Ravi Shanker, 1996). These are medium enthalpy resources, which can be utilized effectively to generate power using binary cycle method.

Since majority of these springs are located in rural India, these springs can support small-scale industries in such areas. Dehydrated vegetables and fruits have a potential export market and India being an agricultural country, this industry is best suited for India conditions.

Map of India showing the geothermal provinces, heat flow values (mW/m^2: in italics) and geothermal gradients (°C/km). I: Himalaya; II: Sohana; III: Cambay; IV: SONATA; V: West Coast; VI: Godavari; VII: Mahanadi. M: Mehmadabad; B: Billimora.

All the geothermal provinces of India are located in areas with high heat flow and geothermal gradients. The heat flow and thermal gradient values vary from 75–468 mW/m^2 and 59–234°C, respectively. Deep Seismic Sounding (DSS) profiles were carried out across several geothermal provinces (Son-Narmada-Tapi; West Coast and Cambay) to understand the crustal structure.

References

1. Akbarali, M.S., Subramanium, S.K., Natarajan, K., Real and Reactive Power Control of SEIG Systems for Supplying Isolated DC Loads. *J. Inst. Eng. India Ser. B*, 99, 587–595, 2018, https://doi.org/10.1007/s40031-018-0350-8.
2. Akbarali, M.S., Subramanium, S.K., Natarajan, K., Modeling, analysis, and control of wind-driven induction generators supplying DC loads under various operating conditions. *Wind Eng.*, pp. 1–16, 2020, https://doi.org/10.1177/0309524X20925398.
3. Yuldoshevb, A., Shoguchkarova, S.K., Kudratov, A.R., Jamolov, T.R., A Study of the Parameters of a Combined Photo-Thermoelectric Installation under Field Conditions. *Appl. Sol. Energy*, 56, 2, 125–130, 2020.
4. Joardder, M.U.H., Halder, P.K., Rahim, M.A., Masud, M.H., Solar Pyrolysis: Converting Waste Into Asset Using Solar Energy, in: *Clean Energy for Sustainable Development*, M.G. Rasul, A. k. Azad, S.C. Sharma (Eds.), pp. 213–235, Academic Press, United Kingdom, 2017, https://doi.org/10.1016/B978-0-12-805423-9.00008-9.
5. Prem Kumar, T., Naveen, C., Premalatha, M., Performance Analysis of 2 in 1 Parabolic Trough Collector for Both Hot Water and Hot Air Production for Domestic Household Applications. *Appl. Sol. Energy*, 55, 397–403, 2019.
6. Qandil, H., Wang, S., Zhao, W., Application-based design of the Fresnel lens solar concentrator. *Renewables: Wind, Water, Solar*, 6, 3, 1–13, 2019.
7. Badal, F.R., Das, P., Sarker, S.K. *et al.*, A survey on control issues in renewable energy integration and microgrid. *Prot. Control Mod. Power Syst.*, 4, 8, 2019.
8. Agarwal, V., Aggarwal, R.K., Patidar, P., Patki, C., A novel scheme for rapid tracking of maximum power point in wind energy generation systems. *IEEE Trans. Energy Convers.*, 25, 1, 228–36, 2010.
9. Bridgwater, A.V., Review of fast pyrolysis of biomass and product upgrading. *Biomass Bioenergy*, 38, 68–94, March 2012.
10. Orozco, R., Redwood, M.D., Yong, P., Cadellari, I., Towards an integrated systems for Bio-energy: Hydrogen production for Escherichia Coil and use of palladium coated waste cells for electricity generation in a fuel cell. *Biotechnol. Lett.*, 32, 12, 1837–45, December 2010.
11. Picarelli, A. and Vargiolu, T., Optimal management of pumped hydroelectric production with state constrained optimal control. *J. Econ. Dyn. Control.* in press, corrected proof Available online 10 June 2020, 103940, vol. 126, pp. 1–24, May 2021.

12. Majorowicz, J. and Grasby, S.E., Deep geothermal energy in Canadian sedimentary basins VS. Fossils based energy we try to replace – Exergy [KJ/KG] compared. *Renewable Energy*, 141, 259–277, October 2019.
13. Pollack, A. and Mukerji, T., Accounting for subsurface uncertainty in enhanced geothermal systems to make more robust techno-economic decisions. *Appl. Energy*, 25415, 113666, November 2019.
14. Van Erdeweghe, S., Van Bael, J., Laenen, B., D'haeseleer, W., Optimal configuration for a low-temperature geothermal CHP plant based on thermoeconomic optimization. *Energy*, 17915, 323–335, July 2019.

2

Present Power Scenario in India

Niraimathi R.[1]*, Pradeep V.[2], Shanthi M.[3] and Kathiresh M.[4]

[1]Department of EEE, Mohamed Sathak Engineering College, Kilakarai, India
[2]Department of EEE, Alagappa Chettiar College of Engineering and Technology, Karaikudi, India
[3]Department of ECE, University College of Engineering, Ramanathapuram, India
[4]Department of EEE, PSG College of Technology, Coimbatore, India

Abstract

The major chunk of power generation in India is done by thermal power plants spread across the nation. These plants are situated near to the coal reserves and near major ports. The working of thermal power plant along with major thermal plants of India is discussed. Indian motherland is blessed with huge potential of hydropower which stands second in producing the highest amount of electric power after coal-based plants. Renewable energy is the fastest-growing in this sector. Solar and wind energy–based power plants are discussed. The promising source for future energy is nuclear power plants. Hence, due importance has been paid to these plants. Specific challenges and opportunities in operating the various power plants are also discussed. India, as a vast land, necessitates bulk power transmission corridors to connect generating stations that are located in close proximity with the sources to the load centres and it is one of the world leaders in this field. This necessitates a discussion of various bulk power transmission lines.

Keywords: Power sector scenario in India, thermal power, gas turbine power plant, hydropower, solar power, wind power

*Corresponding author: rniraimathi27@gmail.com

M. Kathiresh A. Mahaboob Subahani and G.R. Kanagachidambaresan (eds.) Integration of Renewable Energy Sources with Smart Grid, (19–36) © 2021 Scrivener Publishing LLC

2.1 Introduction

Electrical power is the fulcrum for leveraging the economies, as most activities of the present civilization like agriculture and manufacturing revolve with it and change the living standard of people.

India, a rapidly growing economy in the world, is now the third largest producer of electric power with the production of 1,643 TWh of electric energy in the year 2019. It deploys diverse ways of generating electrical energy. Although Indian power sector is dominated by fossil fuels, the Government of India with a view of sustainable development set an ambitious target of renewable installed capacity of 175 GW by FY22. The total installed capacity as of 21 July 2020 is 371,654.13 MW. The thermal generation accounts for 62.29% as shown in Figure 2.1 [1, 2]. After thermal generation, hydropower generation stands second in the list. Wind power and solar power generation are two main renewable energy sources exploited in India [3].

2.2 Thermal Power Plant

In India, thermal energy is the major source of power generation. The thermal power plant produces more than 60% of the electrical power in India.

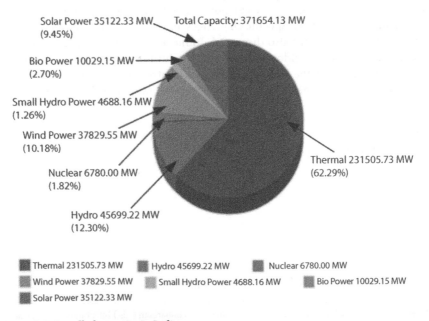

Figure 2.1 Installed capacity in India.

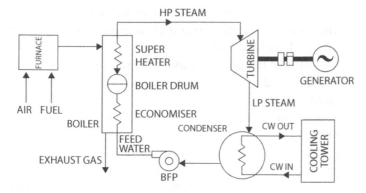

Figure 2.2 A typical thermal power plant.

India is the fifth largest producer of coal in the world [3]. Generally, bituminous coal is used as a boiler fuel in India. Figure 2.2 shows the schematic diagram of a typical thermal power plant.

The heat energy from the combustion of fossil fuels is utilized by the boilers to produce steam at high pressure and temperature. The steam produced is used to drive the steam turbines or steam engines coupled to generators, and thus, electrical energy is generated in the thermal power plant. Steam turbines act not only as prime movers but also as drivers for auxiliary equipment such as pumps and stokers fans [4].

The thermal power plant can be classified into two types:

(i) *Condensing type*: In this type, the exhaust steam is discharged into a condenser, which creates suction at very low pressure and allows the expansion of steam in the turbine to very low pressure and thus increases the efficiency. During this process, steam is condensed into the water which can be re-circulated to the boiler with the help of pumps.

(ii) *Non-condensing type*: In this type, the steam exhausted from the turbine is discharged either at atmospheric pressure or pressure greater than atmospheric. In this type of plant, a continuous supply of fresh feed water is required.

2.2.1 Components of Thermal Power Plant

(i) *Boiler and Boiler Furnace*: A boiler incorporates a furnace to burn the fossil fuel (coal, gas, waste, etc.) and generate heat/steam which is transferred to water to make steam. The types of boiler used in a thermal power plant are water tube and fire tube boiler.

(ii) *Superheater*: It is used to convert the wet steam or saturated steam in to dry steam or superheated steam. Superheated steam contains more heat than the saturated steam at the same pressure. The more heat provides more energy to the turbine hence power output is more.

(iii) *Economizer*: It is used to capture the waste heat from flue gas and transfer it to the boiler feedwater. Economizer alone can produce 10%–12% efficiency increase; thereby, it saves 5%–15% fuel consumption.

(iv) *Condenser*: It is used to convert the steam exhausted from the turbine into the water so that it can be reused again in the boiler. There are two types of condensers are used and they are jet and surface condensers.

(v) *Feed Water Heaters*: A feedwater heater is used in a conventional power plant to preheat boiler feedwater. The source of heat is steam bled from the turbines, and the objective is to improve the thermodynamic efficiency of the cycle.

(vi) *Air Preheaters*: After leaving the economizer, further heat from the flue gas is used to heat the incoming air using air preheater. Flue gas should not be cooled below the dew point to avoid corrosion of the air preheater.

(vii) *Turbines*: There are two types of turbines used:

Impulse turbine: In this turbine, moving and fixed blades are used. The moving blades are mounted on the shaft and fixed blades are fixed to the casing of the turbine. Steam is passed through the fixed nozzles. The steam with high velocity comes out of the nozzle and impinges on the rotor blades.

Reaction turbine: In this turbine, only fixed and moving blades are used. When high-pressure steam passes through fixed blades, then steam pressure drops down, and the velocity of steam increases.

(viii) *Cooling towers*: The cooling tower transfers the heat from the water's heat to the air by directly or by evaporation of the water. So that water coming out of the condenser is reused.

Advantages of Thermal Power Plant

- Fuel cost is comparatively low.
- Installation requires less land compared to a hydropower plant.
- The thermal energy production mechanism is simple and easy.
- The initial cost is lesser compared to other power plants.
- Easy maintenance.

Disadvantages

- The huge production of CO_2 which causes pollution.
- Overall efficiency is less than 30%.
- A huge amount of water is required.
- Warm water comes out of the thermal power plant affects aquatic life.
- Thermal engines require a huge amount of lubricating oil.

2.2.2   Major Thermal Power Plants in India

1. Sasan Ultra Mega Power Plant, Madhya Pradesh: It holds an installed capacity of 3,960 MW situated in Sasan Village of the Singrauli district. Reliance power owns this power plant integrated with a coal mine.
2. Tiroda Thermal Power Plant, Maharashtra: It holds an installed capacity of 3,300-MW plant which covers an area of 454.8 ha owned by Adani power. It uses water from the Wainganga River to go ahead with its operations.
3. Talcher Super Thermal Power Station, Odisha: It holds an installed capacity of 3,000 MW owned by NTPC situated in the Angul district of Odisha.
4. Vindhyachal Thermal Power Station, Madhya Pradesh: It holds an installed capacity of 4,760 MW owned by NTPC. Presently, this is the biggest thermal power plant in India.
5. Mundra Thermal Power Station, Gujarat: It holds an installed capacity of 4,620 MW owned by Adani power. Presently, this is the second biggest thermal power plant in India.
6. Mundra Ultra Mega Power Plant, Gujarat: This is another one situated in Kutch district holds an installed capacity of 4,000 MW owned by CGPL.

7. Rihand Thermal Power Station, Uttar Pradesh: It holds an installed capacity of 3,000 MW owned by NTPC. The plant generates and supplies electricity to different states in the northern part of India such as Uttarpradesh, Rajasthan, Delhi, Punjab, Haryana, and Himachal Pradesh.
8. Sipat Thermal Power Plant, Chhattisgarh: It holds an installed capacity of 2,980 MW owned by NTPC. This is the eighth largest thermal power plant in India.
9. Chandrapur Super Thermal Power Station, Maharashtra: It holds an installed capacity of 2,920 MW operated by Maharashtra state power generation company.
10. NTPC Dadri, Uttar Pradesh: It holds an installed capacity of 2,637 MW owned by NTPC. This is the sixth largest thermal power plant in India.

2.3   Gas-Based Power Generation

India's power generation is contributed almost by all types of conventional and non-conventional power generating plants. But, it is always inclined toward the coal-based power generation. With the increasing new establishments and demands, it is stated that the natural availability of coal will be limited and the coal sector will face a shortfall in the upcoming future.

Due to the given limitations on coal, natural gas plays an important role in India's power sector. As compared to coal, its environmental impacts are comparatively less, which makes gas-based power generation more attractive [5].

2.3.1   Basics of Gas-Based Power Generation

Gas-based power plant is similar to a steam turbine power plant, with the only difference that, here, we use compressed high-pressure air to rotate the turbine. The schematic diagram of a typical gas turbine power plant is shown in Figure 2.3. It has three major parts:

1. Compressor
2. Combustor
3. Turbine

The air or natural gas is compressed in the compressor. The compressed air then passes through the combustor where the air is combusted to high temperature and high pressure. This high-pressure air is used to run the

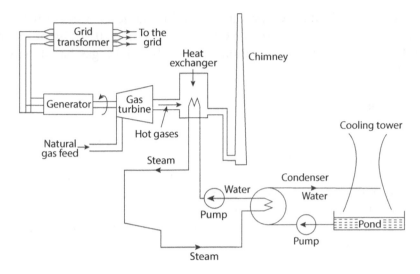

Figure 2.3 A typical gas turbine power plant.

turbine, which, in turn, rotates the alternator to produce electricity. This is a simple cycle gas plant. In a combined cycle gas plant, the heat produced from combustion is used for industrial purposes through a heat recovery steam generator [6, 7].

2.3.2   Major Gas-Based Power Plants in India

Out of 65 gas-based power plants in India, 7 major power plants are owned by NTPC and they are given below:

S. no.	Names	Commissioned capacity (in MW)
1	Anta	419.33
2	Auraiya	663.36
3	Kawas	656.20
4	Dadri	829.78
5	Thaner-Gandhar	657.39
6	Rajiv Gandhi CPP Kayamkulam	359.58
7	Faridabad	431.59

India is the third largest producer and third largest consumer with an installed grid capacity of 370.106 GW as of 31 March 2020. Out of which 48,497 MW of power generation is contributed by gas-based power plants.

Advantages

- As compared with the thermal power plant, it requires small land availability for the power plant setup.
- The construction of a gas-based power plant is simpler.
- Operational cost and the per-unit cost is low.
- It offers black start facilities.
- Burning of natural gases produces less pollutants like NOX and SOX.

Disadvantages

- Efficiency is less.
- Due to high temperature of the system, the lifetime of the power plant is reduced considerably.

The Government of India plans a new scheme to revive a 24,000-MW gas power plant. This proposed scheme hopes to help operate the power stations at 90% capacity by selling the bundled power output with solar energy [2].

2.4   Nuclear Power Plants

Ever since nuclear power came into existence, it has led to various debates on disposal of its wastes and the detrimental effect of such wastes to the mankind. But, many of the nuclear power's advantages are behind closed doors. It has emerged as one of those power-producing methods that produce considerably less amount of greenhouse gases. Enrico Fermi and his team were the people behind today's nuclear power plants. Atoms have loads of energy in them. They only need to be explored by either fission or fusion. Nuclear power plants involve nuclear reactors that contain and control nuclear chain reactions. They also have fuel rods to feed reactors and moderators, control rods, coolants, etc., hope that showed a spotlight on nuclear power [8].

2.4.1   India's Hold in Nuclear Power

It is a prominent fact that India stands third among the world's consumers and producers of electricity. The nuclear plants deliver 6.7 GW of power. Of the total power produced in India, about 2% is produced from Nuclear power. When it comes to worldwide production, India takes 1.2% share, thereby standing at 15th position. India has also signed agreements and treaties with countries like France and Russia to empower and strengthen the nation's nuclear power production [9]. As many plants have been planned and few are already under construction, the position of India and the contribution of India in nuclear power will ameliorate in the impending years. The structure of nuclear power is shown in Figure 2.4.

2.4.2   Major Nuclear Power Plants

There are seven nuclear power plants comprising of 22 nuclear reactors in total operating in India as of March 2018. Of all, Kudankulam Nuclear Power Plant located in Tamil Nadu has the highest capacity power plant with a capacity of 2,000 MW currently under operation and 2,000 MW under operation. It is also the only nuclear plant in India that has incorporated pressurized water reactors while others have boiling water reactors or pressurized heavy-water reactors. The given table gives insights about the major nuclear power plants in India [9].

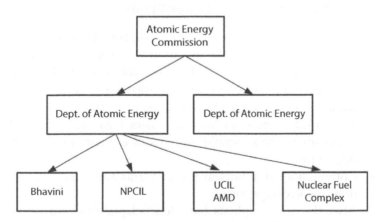

Figure 2.4 Organization chart of nuclear power in India.

2.4.3   Currently Operational Nuclear Power Plants

Power station	State	Total capacity (MW)
Kaiga	Karnataka	880
Kakrapar	Gujarat	440
Kudankulam	Tamil Nadu	2,000
Madras (Kalpakkam)	Tamil Nadu	440
Narora	Uttar Pradesh	440
Rajasthan	Rajasthan	1,080
Tarapur	Maharashtra	1,400

Advantages of Nuclear Power Plants

- The culmination of nuclear power plant is due to its march ahead in reducing greenhouse gas emissions, thereby standing out as a safer way of producing power with less or almost nil release of polluting gases and refraining from contributing to global warming.
- Yet another spotlight in nuclear power is the remarkably low fuel costs and minimal requirement of fuel.
- Nuclear power plants being sustainable and possessing exceptional lifetime.

2.4.4   Challenges of Nuclear Power Plants

- Despite having so many striking pros, the major drawback that blindfolds all of the advantages is the generation of radioactive wastes and challenges involved in its disposal. The disposal is expensive and the wastes are viable of producing ill-effects even after very long periods.
- The harmful effects of nuclear wastes have led to lack of discernment from people and instilled their minds with negativity.
- Commissioning of nuclear power plants and safety precautions involve huge costs.
- Also, the nuclear plants have less adaptability and hence cannot immediately react to changes in the requirement of power.

2.5   Hydropower Generation

Hydropower generation is one of the oldest ways of power generation [10]. With an installed capacity of 4,600 MW (as per March 2020) which accounts for a total of 12.3% total installed capacity in India holds fifth place in terms of exploitable hydro-potential on the global scenario.

Traditionally, potential energy stored in dammed is converted into electrical energy. Based on the height of water, three types of turbines are used. It is capable of instant start. It can take the load at a rate of 20 MW/minute. Therefore, it can be used to meet out peak load periods.

Three gauges project in China on the Yang-Yang river is the largest power station in the world having installed capacity of around 18,200 MW. Tehri hydropower complex is built across the Bagirathi River in Uttarakhand with a proposed capacity of 2,400 MW is the tallest hydropower project in India. The hydropower plants of less than 25 MW come under Ministry of New and Renewable energy and above 25 MW are under the Ministry of Power. The construction of massive dams submerges large areas that may lead to the color change of the land. Some hydro project submerges rain forests which will add to global warming impacts. Because of heavy construction requirements, it needs high capital cost and long gestation period compared to thermal plants [11].

2.5.1   Pumped Storage Plants

The pumped storage plants are used to store hydro electrical energy. This scheme has an upper and lower reservoir. During peak power generation, electrical energy is used to pump water from the lower reservoir to the upper reservoir. During peak load period, energy in the upper reservoir is used to generate the power.

The western region of India with steep gradients of rivers has the largest potential for the development of pumped storage plants. A potential of 96,524 MW was identified by CEA during studies conducted in 1978–1987.

Name of the station	State	Installed capacity (MW)
Srisailam LBPH	Telangana	900
Purulia PSS	West Bengal	900
Kadamparai	Tamil Nadu	400
Ghatghar	Maharashtra	250
Bhira	Maharashtra	150

Synchronous machines in these plants can be made to act as synchronous condensers for voltage support. It helps in the straightening of load curve by act load during low load period and as generating source during peak periods which improves the overall economy of power system operation.

2.6   Solar Power

As we know, the sun is the biggest source of renewable energy for the Earth. The part of solar energy is enormously large even though the earth receives only a part of the energy generated by the sun (i.e., solar energy). There are two forms of solar energy, namely, light and heat, which the earth receives. The solar power can be converted to electrical energy using either photovoltaic (PV) or concentrated solar power systems.

2.6.1   Photovoltaic

PV directly converts solar energy into electricity. When certain semiconductor materials are exposed to light, they absorb photons and release free electrons. This phenomenon is called the photoelectric effect. Direct current (DC) is produced by the PV effect based on the principle of photoelectric effect [12].

Solar cells or PV cells convert sunlight into DC electricity. But a large amount of electricity cannot be produced by single PV cell. Therefore, several PV cells are electrically connected to form a PV module or solar panel. Solar panels range from several hundred watts up to a few kilowatts and are also available in different sizes. Solar panels or modules produce current which is directly dependent on the incident light. Since most of the applications require AC power, the solar power system consists of an inverter too [13].

2.6.2   Photovoltaic Solar Power System

There are different types of PV systems:

- PV direct system: This is the simplest type of PV system. Because they do not use batteries and not tied up to the grid, they power the load only when the sun is shining.
- Off-grid system: This is also called as stand-alone systems which is independent of the power grid. When the sun is not available, batteries are used to store energy.

- Grid connected system: Here solar PV systems are tied with grids so that the excess required power can be accessed from the grid. They may or may not be backed up by batteries.

2.6.3   Concentrated Solar Power System

In this type of solar power system, sun rays are focused on a small area by using mirrors or lenses. Due to this, a large amount of heat is generated at the focused area. This generated heat is used to heat the working fluid which, in turn, drives the steam turbine. There are three main types of concentrated solar power systems, namely, parabolic trough, power tower, and dish sterling system.

Advantages

- Solar power causes no pollution.
- Renewable energy resource and can produce clean power throughout the year.
- Return on investment unlike paying for utility bills.
- In case of grid connected excess power can be sold back to power company.
- Helps the economy by creating jobs for solar manufacturers and installers.
- Less safety risks.

Disadvantages

- Initial cost is high for installation and material.
- Requires additional equipment like inverter to convert dc to ac.
- Efficiency is generally low.
- Cloudy days do not produce as much energy.
- Size of the solar panels varies depending on the geographical locations for the same power generation.
- Large battery bank is needed to store the power.

Government Initiatives: Since solar energy has started to light up the lives of millions of peoples. Government has taken various initiatives to enable an increase in solar power at a subsidized energy cost. There are some top government initiatives which is listed below:

- Solar park scheme [13].
- Viability Gap Funding scheme (VGF).

- Government Yojana solar energy subsidy scheme.
- Ujjwal Discom Assurance Yojna scheme (UDAY) [13].
- Rooftop scheme.
- Jawaharlal Nehru national solar mission.

2.6.4   Major Solar Parks in India

1. Kurnool Ultra Mega Solar Park: It covers a total area of 5,932.32 acres in the Gani and sakunala villages and utilizes 4 million solar panels with capacities of 315 and 320 Watts. This solar park is able to generate more than 8 million KWh of electricity, which is enough to meet the total demand in Kurnool district.
2. Kamuthi Solar Power Park: It is the world's sixth largest solar park and covers a total area of 2,500 acres in Kamuthi, Ramanathapuram district. This solar park is able to generate 648 MW at a single point [14].
3. Gujarat Solar Park: It covers a total area of 5,384 acres in the Charanka village, Patan district and installed capacity is 274 MW. It has also capacity to generate 100 MW of wind power. So, this is the biggest solar-wind hybrid park in the world.
4. Pavagoda Solar Park: It covers a total area of 13,000 acres in Pavagada Taluk, Tumkur district and planned to have a capacity of 2,000 MW at the end of 2018 and will be the world's biggest solar park.
5. Punjab Solar Park: It is located in Bhatinda village, Punjab and installed capacity is 5 MW. The new concept of solar plus organic farming was introduced in this park.
6. Ujjain Solar Park: It is located in Ujjain, Madhya Pradesh and can able to generate 75 MW of solar energy [14]. Panel cleaning is done through the rain water harvesting.
7. Uttarakhand Solar Park: It is located in Maheswari village, Haridwar district and can able to generate 5.5 MW.
8. Ananthapuram Ultra Mega Solar Park: It covers a total area of 7,924.7 acres in Nambula Pulakunta, Ananthapuram district and has a capacity to generate 750 MW.

2.7   Wind Energy

Mankind has always been fascinated by the power of wind. Wind energy is one of the most exploitable forms of energy. From time immemorial, wind power is used for mechanical applications such as water pumping and milling. Professor James Blyth was known to be the first person to successfully produce electricity from wind power in 1887. As of 2019, the total installed capacity of wind power has grown to 651 GW [15].

The kinetic energy in the wind is used to rotate the blades of the rotor which is then used to rotate electrical generators to produce electrical power. The power contained in the wind is proportional to the cube of wind velocity. Therefore, if the wind speed is reduced by 2 percent then there will be an 8% reduction in the power output of the windmill. The place with strong wind is ideal for wind power plants. Generally, the wind speeds increase in fraction as the height increases. Not all the power in the windmill can be converted in to electrical power. Wind turbines are usually place at height of 400 ft. Wind energy is a low quality energy. Theoretically, one extracts a maximum of 16/27th of the total power contained in the wind irrespective of type of wind turbine. This limit is called the Betz limit or Betz coefficient.

Individual wind turbines have an average rating of 3 MW. Many individual wind turbines are connected to form a wind farm. A farm may contain few wind turbines to a large number of wind turbines built over a vast area both onshore as well offshore. Gansu wind farm of china is the largest of its kind in the world. India with an installed capacity of 37.5 GW stands fifth in the world wind power and second in the Asian countries.

Muppandal wind farm spread over Kanyakumari district of Tamil Nadu is the second largest wind farm in the world with 3,000 wind turbines to generate nearly 1,500 MW of electrical power. Tamil Nadu is the leader in wind power installation in India with installed capacity of 7,633 MW which account for 29% of India's total wind energy installations. Gujarat has the highest wind energy potential of 142.6 GW.

The Government of India is keen in enhancing the wind energy sector. National Institute of Wind Energy is established under the aegis of Ministry of New and Renewable Energy sources which carry out activities like wind energy resource, testing, and certification.

Like solar, wind is a highly intermittent source of energy. This poses challenges in grid operation. Mostly, wind energy sites are located far away from the load site which necessitated the establishment of a complex transmission network. The tall structure is visually blocking the aesthetic of the area. Noise pollution is also caused by its blades.

Adequate investment in capacity and efficient working of transmission and distribution systems in developing economies with high growth of electricity demand are important objectives. Market oriented reform processes are required both for the creation of capacity and for electricity as a product. This invariably requires unbundling of transmission and distribution capacities from generation capacities. The importance of such systems for interregional grids across national borders and the superiority of rule-based systems as compared to Bismarckian diplomatic negotiations needs exploration.

Problems of technical management of efficient transmission and distribution systems and in particular of integrating decentralized generation through mini hydel, wind, or PV sustainable generation mechanisms with grids are of interest. In addition to the national level, the integration of such systems with reform at the global level, including the OECD and G20 requires to be explored to integrate state-of-the-art practice in the reform process.

2.8   The Inherited Structure

The inherited model of energy development and consumption in India was that of a centrally planned system in largely publicly owned and operated electricity and energy system, with a substantial Central Government capacity and the rest in State Electricity Boards (SEBs).

The substantial increase in the growth rate of the economy, thermal capacity growth was only 38% of the decade before privatization. Since thermal capacity was a large part of the total the growth of total generation capacity in the Nineties was around half that of the previous decade. The situation in this decade up to 06/07 with a growth rate of 27.31% was worse. In this decade growth of hydel and wind capacity goes up substantially, since it was 33.51% in the decade of the nineties and is already at 38.25% until 06/07. This is particularly true of wind capacity of which growth is encouraging and which stands at 9,000 MW now.

Advances in transmission technologies in India including a large HVDC system have been noted. Technical losses of interregional transfer of power are low and globally comparable. More recently, Power Grid Corporation of India Ltd (PGCIL) has established a 1,200-kV National Test Station for developing the technology for transmitting power at 1,200 kV.

References

1. IEA, *World Energy Outlook 2019*, IEA, Paris, 2019, https://www.iea.org/reports/world-energy-outlook-2019.
2. https://www.energy.gov/science-innovation/energy-sources

3. Khare, V., Nema, S., Baredar, P., Solar–wind hybrid renewable energy system: A review. *Renew. Sustain. Energy Rev.*, Elsevier, 58, C, 23–33, 2016.

4. Elliott, T.C., Chen, K., Swanekamp, R., *Standard Handbook of Powerplant Engineering*, 2nd ed., McGraw-Hill, Europe, 1997.

5. Kehlhofer, R., Hannemann, F., Stirnimann, F., Rukes, B., *Combined-Cycle Gas & Stream Turbine Power Plants*, PennWell Books, Tulsa, Oklahoma, 2009.

6. Zohuri, B., Gas Turbine Working Principles, in: *Combined Cycle Driven Efficiency for Next Generation Nuclear Power Plants*, pp. 147–171, 2015.

7. Walsh, P.P. and Fletcher, P., *Gas Turbine Performance*, 2nd ed., p. 25, John Wiley and Sons, Derby, U.K., 2004.

8. Pioro, I. and Duffey, R., Current status of electricity generation in the world and future of nuclear power industry, in: *Managing Global Warming*, pp. 67–114, 2019.

9. http://www.dae.gov.in

10. *History of Hydropower*, Department of Energy, energy.gov. Retrieved 4 May 2017, https://www.energy.gov/eere/water/history-hydropower.

11. *Status of Hydro Electric Potential Development in India*, Retrieved 17 April 2016, https://cea.nic.in/hepr-report/?lang=enn, Central Electricity Authority, Hydro Electric Potential Reassesment Reports, 2020.

12. Tiwari, G., Tiwari, A., Shyam, *Handbook of Solar Energy*, Springer, Singapore, 2016.

13. Maideen Abdhulkader Jeylani, A., Kanakaraj, J., Mahaboob Subahani, A., Rameshkumar, K., Analysis of Solar PV Application Based on Bidirectional Inverter. *MATEC Web Conf.*, 2018.

14. https://mnre.gov.in/solar/schemes

15. Dawn, S. *et al.*, Wind power: Existing status, achievements and government's initiative towards renewable power dominating India. *Energy Stragey Rev.*, 23, pp. 178–199, 2019, https://doi.org/10.1016/j.esr.2019.01.002Get.

3

Introduction to Smart Grid

G. R. Hemanth[1]*, S. Charles Raja[2] and P. Venkatesh[2]

[1]Department of Electrical and Electronics Engineering, PSG Institute of Technology and Applied Research, Coimbatore, India
[2]Department of Electrical and Electronics Engineering, Thiagarajar College of Engineering, Madurai, India

Abstract

A smart grid (SG) is one of the most developing technologies in the 21st century. In recent years in India, excessive importance has been given for the development and up-gradation of the existing electric grid since the evolving current electric grid into the SG results in benefits for both utility as well as customers, thereby making the grid more reliable and secure. With the increasing number of renewable energy sources (RESs), a lot of challenges have been imposed to balance the electric grid. We have separate networks for the transmission system and distribution system of the SG with Supervisory Control and Data Acquisition (SCADA) systems playing a vital role. A drastic change has occurred by replacing electromechanical and electromagnetic induction type energy meters with smart meters since smart meters not only record data but also analyze the data and communicates with the utility, helping the utility to study the customer load behavior better. This has led to the evolution of a typical architecture for Advanced Metering Infrastructure (AMI) which has been one of the major supports for the SG. In this chapter, the present power generation in India, the need for SG, components of SG at both transmission level and distribution level, the case study of the Puducherry SG pilot project, and the recent trends in SG are discussed in detail.

Keywords: Smart grid, advanced metering infrastructure, energy management system, renewable energy sources, tampering, outage management system, distribution transformer monitoring system, peak load management

*Corresponding author: hemanthgr1994@gmail.com

M. Kathiresh A. Mahaboob Subahani and G.R. Kanagachidambaresan (eds.) Integration of Renewable Energy Sources with Smart Grid, (37–76) © 2021 Scrivener Publishing LLC

3.1 Need for Smart Grid in India

India is the second most populous country in the world with a population of 1.221 billion (Source: Provisional Population Totals: Census 2011, Ministry of Home Affairs, 2011). The installed capacity is 290 GW in 2016 but the electricity demand is expected to be around 900 GW by 2032. Now, India has nearly doubled power generation capacity when compared to the previous decade. However, to meet energy demand in the future, we need a multi-pronged approach. Till now, thermal power plants help to cater to most of the needs of India's electricity generation, transmission, and distribution. Traditional, i.e., conventional, grid has satisfied the user electricity requirement till the 20th century but from the start of the 21st century, it has started to suffer a few limitations. Thermal power plants contribute to air pollution to a larger extent.

The transmission loss is very high since power stations are located at a very large distance from the end consumer. Also, the Aggregate Technical and Commercial (AT&C) losses have become quite high. The key feature in the traditional grid is that there are only a few sources of power generation to meet the huge electricity demand. Management of a power grid with lower losses and better service for the customers will result in the development of the economy [1]. A promising solution for the above problems is renewable energy since it contributes to the distributed power generation. This will help to minimize the transmission loss concerning power transmission and AT&C loss concerning the power distribution. Hence, we need to develop the existing grid and upgrade the design to satisfy the increasing need for energy demand providing opportunities for integration of renewable and conventional energy resources.

Hence, the concept of the smart grid (SG) has been evolved and now is being used almost all over the world to meet our requirements. In our country, during August 2013, a vision document titled "Smart Grid Vision and Roadmap for India" was formulated and released by the India Smart Grid Forum (ISGF) under the Ministry of Power that served as the road map for establishing SGs by 2027 with the main objective of providing 24x7 electricity for all consumers and increasing the energy efficiency [1].

3.2 Present Power Scenario in India

In India, thermal power plants, nuclear power plants, hydropower plants, and renewable energy sources (RESs) are the sources of power generation. Under RES, we have Small Hydro Project, Biomass Gasifier, Biomass Power,

Wind Energy, Solar Energy, Industrial, and Urban Waste Power. The total installed capacity is 367,281 MW [2], and it is shown in Figure 3.1.

It is clear from Figure 3.1 that the private sector holds the major share (nearly 47%) of total installed capacity followed by the state sector (nearly 28%) and the central sector (nearly 25%). This was mainly possible by the deregulation of electricity brought back in 2003 allowing private parties to establish their power generation plants. Further, the distribution of thermal power plants, hydropower plants, nuclear power plants, and RES in the installed capacity is depicted in Figure 3.2. It is observed that thermal power plants have the maximum share in the installed capacity and the distribution of thermal power stations alone is illustrated in Figure 3.3. The coal-fired plants comprise the maximum capacity in thermal power stations.

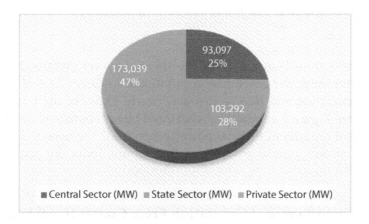

Figure 3.1 Installed generation capacity (in MW).

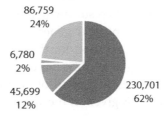

Figure 3.2 Distribution of installed generation capacity.

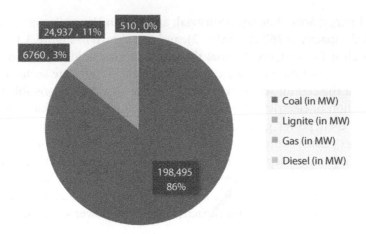

Figure 3.3 Distribution of total thermal power generation capacity.

It is clear from Figure 3.2 that the thermal power plant is the major contributor for power generation accounting for nearly 62.8% of the total power generation and RES contributes nearly 23.5% of the total power generation. Figure 3.3 shows that coal-fired plants contribute nearly 86% of the total capacity of thermal power generation. The overall details for installed power generation stations throughout India are provided in Table 3.1.

3.2.1   Performance of Generation From Conventional Sources

In 2018–2019, the actual conventional generation was 1249.337 trillion units (or) 1249.337 billion units (BU) while the target for the next year 2019–2020 is at 1330 trillion units resulting in a 6.46% increase. Similarly, in 2017–2018, the actual conventional generation was 1206.306 trillion units while the conventional generation for the next year 2018–2019 is at 1249.337 trillion units resulting in a 3.57% increase. This clearly shows that there is a significant percentage of growth of energy generation year by year in the last decade [3] and is given in the Table 3.2.

3.2.2   Status of Renewable Energy Sources

With RESs contributing to nearly 24% of total power generation, they are providing huge scope for alternate power generation. Solar energy has a greater potential in India since in a year, on average, there are 300 solar days. The governments are providing subsidies and other opportunities for

Table 3.1 Installed generation capacity (in MW) of power stations in India.

Region	Ownership/Sector	Mode-wise breakup (MW)			Nuclear	Hydro	RES* (MNRE)	Grand total (MW)
		Thermal						
Northern Region	State	16209.00	250.00	2879.20	0.00	5777.25	701.01	25816.46
	Private	22425.83	1080.00	558.00	0.00	2817.00	15681.74	42562.57
	Central	14354.96	250.00	2344.06	1620.00	11491.52	379.00	30439.54
	Sub Total	52989.79	1580.00	5781.26	1620.00	20085.77	16761.75	98818.57
Western Region	State	21740.00	1040.00	2849.82	0.00	5446.50	555.54	31631.86
	Private	32847.17	500.00	4676.00	0.00	481.00	24560.45	63064.62
	Central	17687.95	0.00	3280.67	1840.00	1695.00	666.30	25169.92
	Sub Total	72275.12	1540.00	10806.49	1840.00	7622.50	25782.29	119866.40
Southern Region	State	19932.50	0.00	791.98	159.96	11774.83	586.88	33246.15
	Private	12747.00	250.00	5340.24	273.70	0.00	41207.02	59817.97
	Central	11835.02	3390.00	359.58	0.00	0.00	541.90	19446.50
	Sub Total	44514.52	3640.00	6491.80	433.66	11774.83	42335.80	112510.62
Eastern Region	State	7450.00	0.00	100.00	0.00	3537.92	275.11	11363.03
	Private	6153.00	0.00	0.00	0.00	96.00	1211.86	7460.86

(Continued)

Table 3.1 Installed generation capacity (in MW) of power stations in India. (*Continued*)

Region	Ownership/Sector	Mode-wise breakup (MW)							RES* (MNRE)	Grand total (MW)
		Thermal					Nuclear	Hydro		
	Central	13812.05	0.00	0.00	0.00	13812.05	0.00	1005.20	10.00	14827.25
	Sub Total	27415.05	0.00	100.00	0.00	27515.05	0.00	4639.12	1496.97	33651.13
North Eastern Region	State	0.00	0.00	497.71	36.00	533.71	0.00	422.00	233.25	1188.95
	Private	0.00	0.00	24.50	0.00	24.50	0.00	0.00	100.95	125.45
	Central	770.02	0.00	1253.60	0.00	2023.62	0.00	1155.00	30.00	3208.62
	Sub Total	770.02	0.00	1775.81	36.00	2581.83	0.00	1577.00	364.20	4523.02
Islands	State	0.00	0.00	0.00	40.05	40.05	0.00	0.00	5.25	45.30
	Private	0.00	0.00	0.00	0.00	0.00	0.00	0.00	7.84	7.84
	Central	0.00	0.00	0.00	0.00	0.00	0.00	0.00	5.10	5.10
	Sub Total	0.00	0.00	0.00	40.05	40.05	0.00	0.00	18.19	58.24
ALL INDIA	State	65331.50	1290.00	7118.71	236.01	73976.21	0.00	26958.50	2357.03	103291.74
	Private	74173.00	1830.00	10598.74	273.70	86875.45	0.00	3394.00	82769.86	173039.30
	Central	58460.00	3640.00	7237.91	0.00	69337.91	6780.00	15346.72	1632.30	93096.93
Total		197964.50	6760.00	24955.36	509.71	230189.57	6780.00	45699.22	86759.19	369427.97

Table 3.2 Generation and growth in conventional generation in India from 2009–2010 to 2019–2020.

Year	Energy generation from conventional sources (BU)	% of growth
2009–2010	771.551	6.6
2010–2011	811.143	5.56
2011–2012	876.887	8.11
2012–2013	912.056	4.01
2013–2014	967.150	6.04
2014–2015	1048.673	8.43
2015–2016	1107.822	5.64
2016–2017	1160.141	4.72
2017–2018	1206.306	3.98
2018–2019	1249.337	3.57
2019–2020*	1054.785	1.10

*Source: Central Electricity Authority (CEA).

users to establish solar-powered installations. The Ministry of Power has set a target that solar energy alone will meet nearly 8% of overall energy demand by 2022.

Figure 3.4 illustrates the installed capacity of RES [2]. In March 2017, approximately, 57.24 GW is the grid-linked renewable energy potential (excluding the major hydropower) out of which 57% (32 GW) of renewable energy comes from the wind while solar contributes nearly 21% (12 GW). Recently, in 2017, the "Power for All" plan was launched by the Indian government with a major focus on renewable energy expansion to 175 GW by 2022. Out of 175 GW, 100 GW is for solar, 60 GW for wind, 10 GW biomass, and 5 GW limited hydropower. This is shown in Figure 3.5.

3.3   Electric Grid

The greatest technological accomplishment of the 20th century is the power grid. A transition has occurred from a centralized generation to a de-centralized generation. Over a century, in centralized power generation, the power plants feed energy over high transmission lines into the grid at

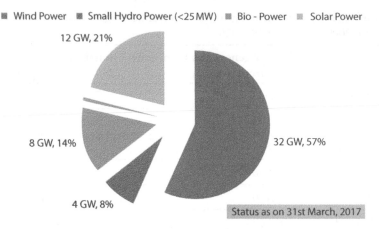

Figure 3.4 Installed generation capacity of renewable energy sources (in GW).

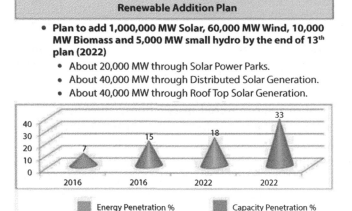

Figure 3.5 Renewable capacity addition plan.

a limited number of points while consumers extract energy from the grid over a low voltage distribution lines at a million points. This indicates unidirectional energy transfer from power producer to power consumer. A shift in paradigm is observed from centralized generation to distributed generation. The picture of the grid is changing since the last decade. Some of the visible characteristics of this shift are as follows [4]:

- The difference between production, transmission, and delivery is greatly reducing.

- The unidirectional power grid has transformed into a bi-directional power grid with the exchange of both energy and information.
- The initial opinion of investing in transmission has now changed to investing in storage especially in regions where transmission and distribution losses are very high.
- The loads have changed from incandescent lamps and induction motors that can handle variations in grid voltage and frequency in the past to digital loads of today that require constant grid frequency and voltage.
- The electricity demand which previously was inflexible has now become dependent on price.
- The balance in grid power now greatly depends on both solar photovoltaic plants and wind power since they are stochastic sources of power generation.
- With numerous electric cars operating, the grid has become much more complex since they serve as virtual power plants that can serve demand on a short-term basis.

Previously, power plant operators had minimal control in the conventional or traditional grid. Now, more emphasis is given to control and automation in the SG since it is fitted with numerous sensors and smart meters that involve both energy and information thus evolving the SG into a grid of things.

3.3.1 Evolving Scenario of the Electric Grid

Some of the disruptive changes taking place in the power systems are described in detail below [4]:

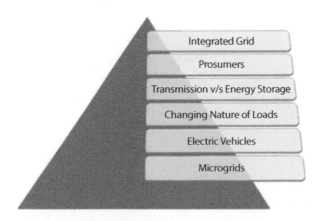

3.3.1.1   Integrated Grid

The traditional power grid is vertically divided into three parts of generation, transmission, and distribution, respectively. The transmission accounted for all generation receipts. Also, the organizational structures and functions are separated for generation, transmission, and distribution. There exists only a flow of energy from generation to distribution. This is shown in Figure 3.6. Now, with an increasing number of modern energy generation such as micro-wind turbines, rooftop solar PV, and battery-based energy storage systems (ESSs), the differences between generation, transmission, and distribution are rapidly decreasing. By increasing the number of installations for distributed generation, it has become essential to predict energy demand in advance.

But now, with distributed energy resources occupying a major share of the energy mix, it is essential to accurately predict the energy demand in advance. This is illustrated in Figure 3.7. This necessitates the development of new design frameworks and operational rules that requires new investment capital. Thus, the installation, operation, and maintenance of the emerging SG pose a major challenge.

3.3.1.2   Prosumers

For over a century, there is only a unidirectional flow of electrons from power generating stations to loads at the consumer's premises. All the operational

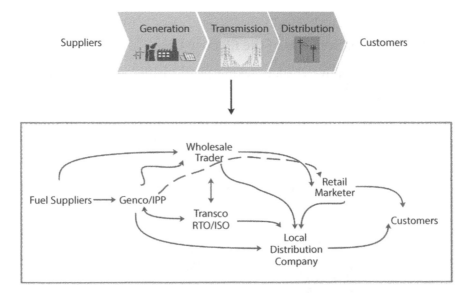

Figure 3.6 Layout of traditional grid.

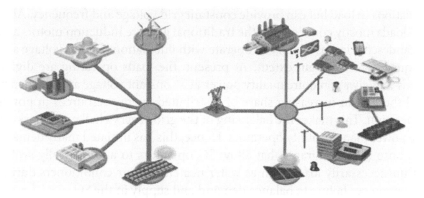

Figure 3.7 Layout of developing integrated smart grid.

structures, methods, and machines were planned based on this one-way energy flow. But now, with the incorporation of renewable energy resources, it is essential to think of new metering infrastructure and utility must be able to detect reverse power flow from customer's premises in the low-voltage grid. Hence, the end consumer who was previously only an energy user has now turned into both a producer and a consumer - "prosumer". The utility must evaluate the last mile network capacity based on load flow studies and network models and plan before connecting rooftop solar PV installations.

3.3.1.3   Transmission v/s Energy Storage

The erection of transmission lines has been a huge challenge in many geographical locations depending on the topography especially if the user is located in a congested area or a hilly location. It usually takes several years to even decades for completing the construction of a transmission network. This further results in an increased cost of construction exceeding the planned amount. A wide range of energy storage solutions is now entering into the grid that includes lithium-ion batteries and sodium sulfide batteries operating on an MW-scale. Now, with the evolving concept of distributed generation, storage gets added by default in the customer location, thus greatly minimizing the network losses when compared to the traditional power generation model.

3.3.1.4   Changing Nature of Loads

Most of the power generation plants in the conventional or traditional grid are thermal power plants and hydro-power plants which are tolerant to variations in load but can provide constant grid voltage and frequency. Also, the loads mostly connected to the traditional grid are induction motors and incandescent lamps which can operate with fluctuations in grid voltage and frequency to a certain extent. At present, the loads operating are digital loads that always require quality power at a constant voltage and frequency and there is an increasing share of distributed energy resources in power generation. This makes the balancing of the grid, a very challenging task for the power plant or utility operators. Hence, this has initiated many demand response (DR) programs that allow SG operators to automatically switch off un-necessarily loads such as water heaters and air conditioners during abrupt power failure to balance demand and supply in the SG.

3.3.1.5   Electric Vehicles

Electric vehicle (EV) is the proposed solution agreed internationally to handle air pollution in majorly affected cities of the world. It is simply practical electrification of transport. In 2013, "National Electric Mobility Mission" was launched by the Government of India (GoI) that had set a target of 20 lakh four-wheelers and 40–50 lakh two-wheelers by 2020. This was initially slow because of the limited number of charging stations. But now, this initiative is highly supported by GoI since it is viewed as a potential solution to tackle air pollution in the highly polluted cities. For grids, EVs are considered as big loads because the size of car batteries ranges from 10 to 100 kWh while the size of bus batteries vary from 100 to 300 kWh.

This has a very serious impact on the distribution grid involving redesigning of transformers and cables need to be done at many points depending on the movement of EVs. Also, a large number of charging stations need to be established that requires a large investment. EV batteries can inject power back into the grid when not in use, and hence, it also serves as the energy storage. This also necessitates a separate tariff for EVs. A large number of EVs connected to the grid are viewed as automated power plants that could help balance demand and supply. Figure 3.8 shows the technologies involved in the EV.

3.3.1.6   Microgrids

One of the latest SG innovations in developing countries is microgrid. The main reason behind this the lack of emergency power supply networks in airports, hospitals, etc. The central part of a microgrid is an autonomous control center that is capable of disconnecting micro grid from the power

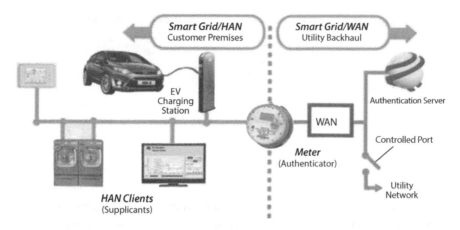

Figure 3.8 Layout and technology used in EV.

grid and monitoring and shedding the load inside the microgrid to match energy demand with generation during emergency condition and also accessing the storage. This is seen as a protective measure against cyber-attacks. With increasing DC generation and DC consumption, we can have DC distribution along with AC distribution. In 2013, ISGF established "Low Voltage Direct Current" (LVDC) Forum which has setup an operating voltage of 48 V DC for indoor applications in India. Rural electrification can be made effective with DC grids.

3.4 Overview of Smart Grids

A SG is the evolution of the traditional grid with sophisticated control, automation, information technology (IT), and the Internet of Things (IoT) that control power flow from power generation to power consumption and monitors the bidirectional energy flow in real time. It is an electric grid that provides numerous choices for consumers to use energy by means of smart appliances that support real-time pricing. In the SG, the bidirectional flow of energy and information/exists from multiple points of generation to multiple points of consumption [4].

3.4.1 Purpose of Smart Grid

As discussed earlier, there is a shift from a centralized generation toward a decentralized generation. The traditional differences between generation, transmission, and distribution are rapidly reducing, and the grid is transforming

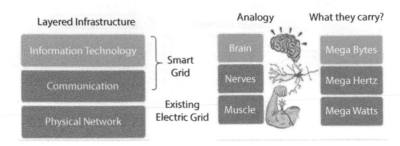

Figure 3.9 Relation of existing grid and smart grid with human body.

into an interconnected SG, a unique solution that comprises various forms of power production and makes the buyer both a producer and a consumer (prosumer). Each house will consume, store as well as generate and sell the power to the grid. The customers take active participation in the DR programs to shift their energy usage patterns from expensive peak hours to cheap off-peak hours. SG technologies enable the availability of quality power.

Figure 3.9 shows the analogy existing between SG and human body. The existing grid is represented by the muscles while the SG is represented by the brain and nerves. The main objective is to make the existing grid smart by employing smart sensors, smart meters, automation, control, IT, and real-time monitoring systems that are capable of providing two-way communication. The main promoters for SG in our Indian context involving all the stakeholders are utilities, customers, and governments and regulators [4].

3.5   Smart Grid Components for Transmission System

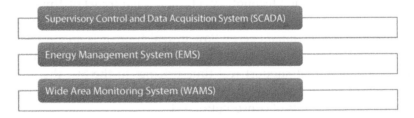

3.5.1   Supervisory Control and Data Acquisition System

The Extra High Voltage (EHV) transmission network operating with a voltage of 110 kV and above is equipped with communication and real-time automation, monitoring, and control systems. So, they are smart or intelligent from the past. The control center also called the load dispatch center of EHV systems controls and monitors power flows in real-time by using Energy Management System (EMS) and Supervisory Control and Data Acquisition (SCADA). The EHV network has effective communication systems between load dispatch or control, center, and each generating station and EHV substation for the effective performance of SCADA/EMS. The operators control loads at the substations as well as generators from the control center [4].

3.5.1.1   SCADA Overview

SCADA refers to a system that is capable of collecting data from various sensors and other devices at a location, i.e., industry, factory, substation, and power station, and sending the data to a central computer. This central computer then controls and monitors the entire system in real-time. The major advantage in SCADA is that it requires minimum human effort since the major role is played by sensors.

3.5.1.2   Components of SCADA

The SCADA comprises of a Master Station and a number of other Remote Stations that are equipped with the following components:

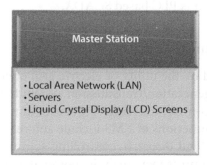
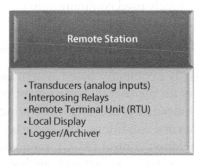

Master Station

• Local Area Network (LAN)
• Servers
• Liquid Crystal Display (LCD) Screens

Remote Station

• Transducers (analog inputs)
• Interposing Relays
• Remote Terminal Unit (RTU)
• Local Display
• Logger/Archiver

SCADA performs different function such as producing alarms, data logging (recording), load monitoring with display and logging, data acquisition, supervisory control, tagging, and time synchronization of Remote Terminal Units (RTUs) and trending.

RTUs are installed at every field equipment and substations. SCADA uses RTUs to communicate with many external devices such as industrial meters and programmable logic controllers (PLCs). This is shown in Figure 3.10. A variable called "tag" is assigned for every data read from the sensors by SCADA. The same is also used for data write operation. The process of

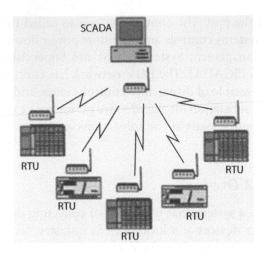

Figure 3.10 Interconnection of RTUs in SCADA.

creating and managing tags is called tagging. A way of selecting, processing, analyzing, evaluating, and displaying data is provided by "trending" analysis. The various communication systems used in SCADA include power line carrier communication, analog communication, fiber optic cables, microwave communication, satellite communication, and public telecom operators. The bandwidth is limited in analog communication. The widely used communication system is PLC based SCADA.

3.5.2   Energy Management System

EMS is a computer-based system used by the utility or electric grid operators for the overall management of generation/transmission system that includes data processing, monitoring, control, and triggering alarms during abnormal conditions. The important functions of EMS include automatic generation control (AGC), economic load dispatch, controlling spinning reserve, real-time network analysis, contingency analysis, power flow or load flow study, grid monitoring, load forecasting, and scheduling transactions. EMS along with SCADA provides efficient and economic operation of the transmission system [4].

3.5.3   Wide-Area Monitoring System

The system used for monitoring real-time should be very fast, accurate, and capable of handling large volumes of data. The measurements that are fast and accurate with high throughput are provided by phasor measurement units (PMUs). Expected latency varies from 100 milliseconds to 5 seconds. PMU provides data in milliseconds while RTU provides data in seconds. RTU data resembles an X-ray while PMU data resembles an MRI scan. This overall system functioning over a wide area using PMU data is called Wide-Area Monitoring System (WAMS) [4].

3.6   Smart Grid Functions Used in Distribution System

The distribution network comprises a medium-voltage network operating at a voltage of 33 kV/11 kV and a low voltage network operating at 415 V/230 V. From the past, the distribution system is provided with less automation. One of the several reasons is the communication cost. When the customers complain about the power outage, then a team is sent to handle the issue and resolve it. At present, one of the SG initiatives is to modernize the distribution system with advanced control, measurement, and automation system [4].

Major Smart Grid Functions Used in Distribution System
• Supervisory Control and Data Acquisition (SCADA)
• Distribution Management Systems (DMS)
• Distribution Automation (DA)
• Substation Automation (SA)
• Advanced Metering Infrastructure (AMI)
• Geographical Information System (GIS)
• Peak Load Management (PLM)
• Demand Response
• Power Quality Management (PQM)
• Outage Management System
• Distribution Transformer Monitoring System
• Application Integration
• Smart Street Lights (with noise and pollution sensors)
• Energy Storage
• Cyber Security
• Analytics

3.6.1   Supervisory Control and Data Acquisition System

The SCADA system used in the transmission system can also be used for the distribution system. Field Remote Terminal Units (FRTUs) are used in distribution substations at the locations of distribution transformers (DTs). The options provided for communication are similar in both the transmission system and distribution system. The choice to be used is decided by the utility or grid operator.

3.6.2   Distribution Management System

The computer-assisted system used for efficient and reliable control and management of the distribution network is termed Distribution Management System (DMS). It is supported by various software applications. The functions of DMS include network planning, network design, network analysis, protection, remedial actions, and load flow studies in the distribution system.

3.6.3   Distribution Automation

Distribution Automation (DA) refers to different automation methods used to enhance the performance of the power distribution system by making use of sensors and other devices to monitor the operating conditions of the grid that include power flow. Now, grid operators can intervene either manually or using an automated control mechanism. DA is considered as an important technique used for outage prevention. It has the ability to interact with the SG ensuring the reliable operation of the SG. It allows renewable energy penetration that facilitates bidirectional power flow utilizing different control mechanisms. The key components of DA include sectionalizer, recloser, fault locator, ring main unit (RMU), and capacitor bank. These are shown in Figure 3.11.

Sectionalizer is a protective device that is used along with relay, and circuit breaker to isolate the faulted part of the distribution system. It however cannot interrupt fault current. Recloser is a device used to operate the circuit breaker. It has the capacity to reclose up to a preset count, thereby interrupting fault current before isolating the breaker from the circuit. Once the preset count is reached, the recloser opens and operates the

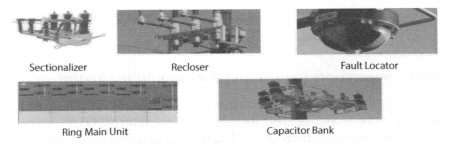

Figure 3.11 Parts of distribution automation system.

breaker. Fault locators are used to identifying the faults occurring in a location. They automatically isolate and re-route the power to the needed areas and alert the concerned personnel to repair the fault. RMUs are installed at suitable points in the feeder to check and monitor the performance of sectionalizer, recloser, and other equipment in the network. Capacitor banks are used to monitor the voltage and maintain the power factor.

3.6.4   Substation Automation

Substation Automation (SA) system refers to a system used by the utility to remotely control and monitor the installed components inside a substation and automatically collect data. The major components of SA include digital relays and communication systems. It is used to restore the service after long interruptions caused by equipment failures, faults, lightning, etc.

3.6.5   Advanced Metering Infrastructure

Advanced Metering Infrastructure (AMI) is considered one of the main pillars of the SG. The bidirectional communication protocols used in AMI include Home Area Network (HAN), Field Area Network (FAN), Wide Area Network (WAN), and Neighborhood Area Network (NAN). Meter Data Management System (MDMS) collects data from several smart meters and sends it to the Head End System (HES) of the utility. This HES communicates with Data Concentrator Unit (DCU), gateway, router, and access point via WAN/Backhaul. Post-processing, the data is sent to the billing system and other IT applications. HAN comprises household appliances like television, air conditioner, water heater, washing machine, etc. The central part of AMI is the smart meter. The architecture of a typical AMI is shown in Figure 3.12. The existing meters are made smart by using a two-way machine-to-machine (M2M) communication, a remote connect switch, and a remote disconnect switch. Smart meters use near real-time sensors, power-failure monitoring devices, and power quality monitoring devices. In-Home Displays (IHDs) are used in customer premises to monitor energy consumption in real time during a DR program. Now, smartphones have many applications that can replace IHDs.

For residential customers, the basic hardware components required are discussed in [6]. The key components involved are the electricity meter, gas meter, water meter, plug-in smart meter, sensors, data logger, local communication hub, and the dashboard or IHD is shown in Figure 3.13.

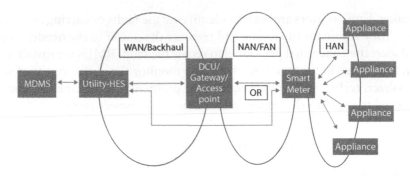

Figure 3.12 Layout of AMI.

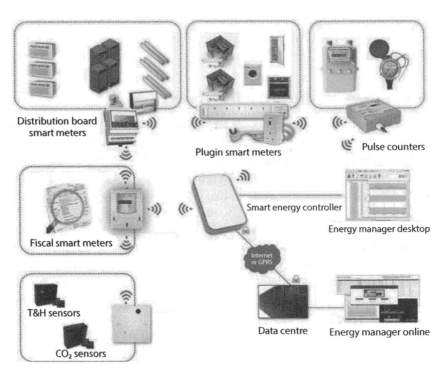

Figure 3.13 Key components of smart metering system.

Electricity, gas, and water meter: They are used to measure the consumption of basic quantities that include electricity, gas, and water. The different types of electricity meters are standard, dial, and digital meters. The communication methods vary depending on the type of electricity meter. For old or conventional meters, the LED type meter was used. In the latest smart meters, the pulse output produced is sent to a data logger.

Plug-in smart socket: It is most suited for domestic, and small business environment. Different appliances such as washing machine, microwave oven, or a personal computer are plugged in this smart socket and it sends data to the data logger once in every 10 seconds for analysis.

Sensors: Intelligent electronic devices (IEDs) are used to sense the environmental parameters such as humidity, temperature, light, and carbon dioxide and send the captured data to a data logger, and further sent to the central server.

Data logger: It receives data from different sensors in the SG and connects with a central server or personal computer and is also equipped with the option of saving data.

Local communication hub: Local Area Network (LAN) and Home Area Network (HAN) are used as a means of communication of various sensors and components with the data logger. Then, the data is transferred to a central processing unit (CPU) and finally sent to the utility operators.

Dashboard or IHD: The data insights are provided for the customers by the IHDs fitted in the residential premises. The data displayed includes the predicted monthly energy consumption, cost/tariff of energy consumption, etc.

3.6.6 Geographical Information System

All the components of the SG are mapped to a digital database called Geographical Information System (GIS). It is a computer-based system used by grid operators to schedule and control their operations and also locate the different components of the SG. The features of GIS are shown in Figure 3.14. It provides information related to weather conditions, SCADA, and customers. GIS maps get automatically updated daily. Whenever an

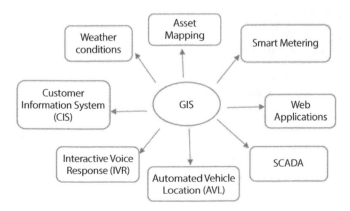

Figure 3.14 Functions of GIS.

asset is added or removed, the information gets periodically updated in the GIS map. It is considered a valuable feature for personnel involved in providing water, gas, and other transport.

3.6.7 Peak Load Management

Peak load management (PLM) consists of all strategies involved in controlling the peak demand such as DR programs, critical peak pricing, and time of use tariffs.

3.6.8 Demand Response

DR is a program that allows the utility to automatically connect or disconnect the loads. This involves active participation from the customers. The customers are rewarded or penalized depending on their actions performed during peak hours and off-peak hours. Based on the mutual agreement, the customers shift their load from peak hours to off-peak hours. The process flow followed in DR program is shown in Figure 3.15.

DR avoids the use of additional generation, transmission, and distribution capacity and also helps to prevent blackouts and unnecessary power outages. Open ADR standard is used to display the Auto Demand Response.

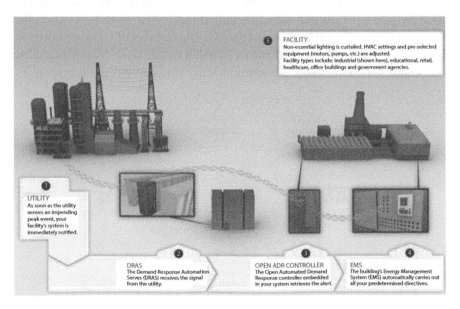

Figure 3.15 Process chart for demand response.

3.6.9   Power Quality Management

The quality of power is decided by the maintaining the voltage within the operating limits. Power quality issues are associated with interruptions and variations in the supply voltage as well as voltage sags and swells. The major power quality problem faced now is the harmonics introduced in our power system by the customer loads such as switched mode power supply (SMPS), uninterrupted power supply (UPS), induction furnace, induction stove, and washing machines. With the increasing renewable energy resources, ensuring the power quality is a challenging task. Ideally, voltage and current waveforms should be sinusoidal. But due to the presence of harmonics, both voltage and current waveforms become square waveforms. These harmonics can be removed using custom power devices such as distribution static compensator (DSTATCOM), dynamic voltage restorer (DVR), and unified power quality conditioner (UPQC).

3.6.10   Outage Management System

The system that is capable of identifying and resolving the outages and reporting data is called the Outage Management System (OMS). During power outages, customer complaints can be resolved using OMS based GIS. The main objectives of OMS are fast identification of power outages and minimizing the response time of customer complaints. It can be afforded at a marginal cost for service providers and public infrastructure developers.

3.6.11   Distribution Transformer Monitoring System

In India, most of the DTs experience overloading and phase imbalance during summer that eventually leads to the burn out of DTs. There is a huge investment needed for the construction of DTs. Hence, monitoring DTs is very important. These DT monitoring systems help to avoid an imbalance of three-phase DTs by loading them equally on all the phases. When the load in the distribution system increases, the overloaded DTs are replaced by the higher capacity DTs.

3.6.12   Enterprise Application Integration

The data collected from different components in the SG are used by various applications that need a middleware framework to provide data integrity. One such middleware framework tool is Service-Oriented Architecture (SOA). The existing database and applications are all linked with the middleware tool so that we can easily retrieve data from the same database depending upon the applications.

3.6.13   Smart Street Lights

The latest technology in developing smart cities is smart street lighting. Conventional sodium vapor lamps used in street light consumes more power. Hence, they are replaced by LED lamps that operate for more light hours and consume less power. These LED lamps are provided with more sophisticated features such as remote on and off control using General Packet Radio Services (GPRS), Wi-Fi, or RF Mesh, increasing and decreasing illumination, etc. The latest smart street lights have embedded noise sensors and emission sensors in the street poles.

3.6.14   Energy Storage

A major role in the SG in maintaining generation demand balance and also the energy requirements of the grid is performed by the ESSs. It helps to meet emergency conditions and increase the quality and reliability of power produced by RES. This is essential for the modernization of the SG. Some of the existing ESSs include different types of batteries, pumped hydro plants, micro-wind turbines, flywheels, and capacitors.

3.6.15   Cyber Security

Cybersecurity deals with the protection of both hardware and software assets from intended, and unintended activities. This is also essential for the protection of the SG from unauthorized access. There is a dependency of physical assets on cyber assets which has made utility operators make SG resilient and robust. In 2014–2015, a set of guidelines for enhancing cybersecurity in the SG was jointly developed by ISGF, National Critical Information Infrastructure Protection Centre (NCIIPC) and Veermata Jijabai Technological Institute (VJTI). VJTI is a famous engineering institution in Mumbai.

3.6.16   Analytics

The analytics is essential for SG operators to analyze the SG data and visualize the load usage patterns for the detection of any abnormalities. It helps to find irregular patterns and other operational parameters such as pricing to make the grid more efficient. This, in turn, helps to reduce the grid operating costs and losses existing in the transmission and distribution, and satisfy the needs of the customers.

3.7   Case Study: Techno-Economic Analysis

3.7.1   Peak Load Shaving and Metering Efficiency

One of the main strategies used in the SG is peak load shaving, i.e., PLM. It refers to the shaving of energy when the load is shifted from peak hours to off-peak hours. Figures 3.16 and 3.17 show the actual and mean hourly load profile of two consumers in SP Nagar and DT [7]. The actual hourly load profile is shown in red color while the mean hourly load profile is shown in green color.

The percentage shavings achieved for three consumers and the DT during each hour of the day are shown in Table 3.3.

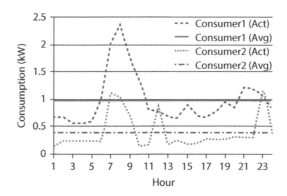

Figure 3.16 Actual and mean hourly consumption profile of consumers 1 and 2.

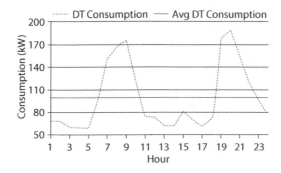

Figure 3.17 Hourly consumption in distribution transformer.

Table 3.3 Energy savings through peak load management.

Hour (of the day)	Actual kWh	Average kWh	Difference (peak-mean)	Shaving
Consumer 1 (Average Shaving of 28.8%)				
6	1.023		0.043	4.20%
7	2.055		1.075	52.31%
8	2.366		1.386	58.58%
9	1.765	0.98	0.785	44.48%
10	1.335		0.355	26.59%
21	1.216		0.236	19.41%
22	1.189		0.209	17.58%
23	1.059		0.079	7.46%
Consumer 2 (Average Shaving of 59.35%)				
7	1.1		0.71	64.55%
8	1.04	0.39	0.65	62.50%
9	0.7		0.31	44.29%
23	1.15		0.76	66.09%
Consumer 3 (Average Shaving of 45%)				
6	4.79551		1.3845	28.87%
7	7.7234		4.3124	55.84%
8	8.05		4.639	57.63%
9	9.0234		5.6124	62.20%
10	7.77409	3.411	4.36309	56.12%
19	3.99172		0.58072	14.55%
21	6.75426		3.34326	49.50%
22	6.29465		2.88365	45.81%
23	5.20742		1.79642	34.50%

(Continued)

Table 3.3 Energy savings through peak load management. (*Continued*)

Hour (of the day)	Actual kWh	Average kWh	Difference (peak-mean)	Shaving
Distribution Transformer (Average Shaving of 34.66%)				
7	150.5		51. 16	33.99%
8	165.8	99.34	66.46	40.08%
9	175.4		76.06	43.36%

From Table 3.3, it is observed that consumers 1, 2, and 3 have an average demand of 0.98, 0.39, and 3.411 kWh, respectively. Consumers 1, 2, and 3 have an average peak load shaving of 28.8%, 59.35%, and 45%. The DT has an average demand of 99.34 kWh and the average peak load shaving of 34.66%. This indicates that PLM is vital for small consumers as well as utility operators. This is essential to reduce the distribution losses and improve system reliability.

3.7.2   Outage Management System

OMS is essential to improve the reliability of the SG. It functions along with AMI, ICT, and DA for improved efficiency and performance of the overall system. Outage control was done recently in the Puducherry pilot project with a tracking system for Fault Passage Indicators (FPIs) and a distribution transformer monitoring system (DTMS). DTMS comprises palm sensors, oil temperature sensors, current transformers (CTs), and potential transformers (PTs) to measure the oil level, temperature, and phase current. The details of DTMS and FPI are given in Tables 3.4 and 3.5, respectively [7].

Figure 3.18 shows the three-phase instantaneous currents. Instantaneous currents of 281 A, 254 A, and 267 A flow through the three phases A, B, and C respectively. Figure 3.19 shows the average current flowing through all the three phases for 30 days from March 18 to April 18, 2014, captured by GridSense. During this period, it ranges between 199 A and 339 A.

Figure 3.20 shows the average daily percentage loading in March 2014. It ranges between 29% and 49%. Figure 3.21 shows the different temperature profiles in DTMS corresponding to oil temperature, enclosure temperature, and winding temperature, respectively. Figure 3.22 depicts an FPI installed in Puducherry SG pilot project. In total, 21 FPIs are installed out of which 6 FPIs are witha communication interface and 15 FPIs are without a communication interface. FPIs can easily access the data using a web link.

Table 3.4 Details of DTMS installed in Puducherry SG pilot project.

Supplier	Sensors		
	Oil temperature	Palm	CT, PT
Sharika	Yes	No	Yes
AMI Tech	Yes	Yes	No
Gridsense	Yes	No	Yes
AMI Tech	Yes	Yes	No
Sai Electronics	Yes	Yes	Yes

Table 3.5 Details of faults detected by FPI during July to November 2014.

Fault	Primary fault	Secondary fault
July	38	8
August	20	8
September	13	4
October	13	5
November	7	1
Total	91	26

Figure 3.23 shows a sample message generated from FPI. It comprises feeder data, fault data, date, and time. Figure 3.24 shows the fault data report of OMS. It comprises the type of fault, latitude, longitude, the current flowing through the conductor, date, and time.

3.7.3 Loss Detection

The smart meters installed in the Puducherry pilot project [7] communicate the collected data to a control server once in 30 minutes. The energy loss and power flows were measured at different nodes and the terminal of DT by software. This software computes the difference between data from DT and cumulative data collected from each smart meter of the customer. This difference is equal to the system losses.

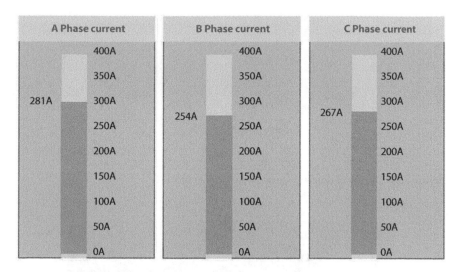

Figure 3.18 Instantaneous current of all three phases of a DT.

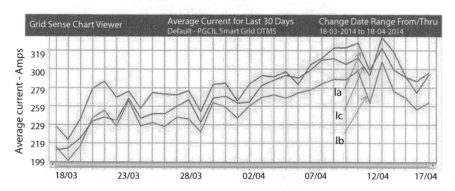

Figure 3.19 Average daily current during March 18 to April 18, 2014.

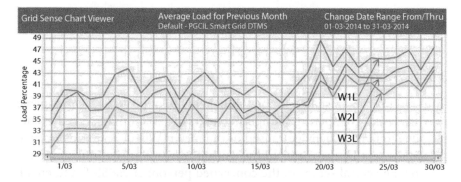

Figure 3.20 Average daily percentage loading in March 2014.

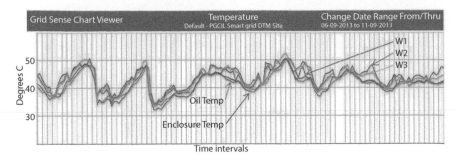

Figure 3.21 Display of different temperature profiles of a DT.

Figure 3.22 FPI Installed in Puducherry SG pilot project.

3.7.4 Tamper Analysis

The smart meters installed in the Puducherry pilot project [7] are capable of withstanding tampering. Tampering refers to variations observed in current due to fault or damage caused in the meter. When tampering occurs, an SMS is sent by the meter to the central server. Further information is sent by the central server to the concerned personnel via SMS and email. Table 3.6 shows the details of eight tampering actions. The major causes

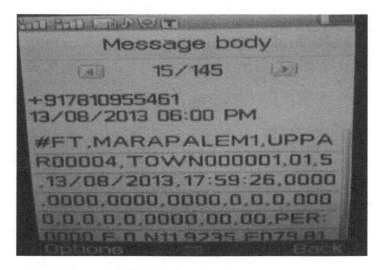

Figure 3.23 Message generated from FPI.

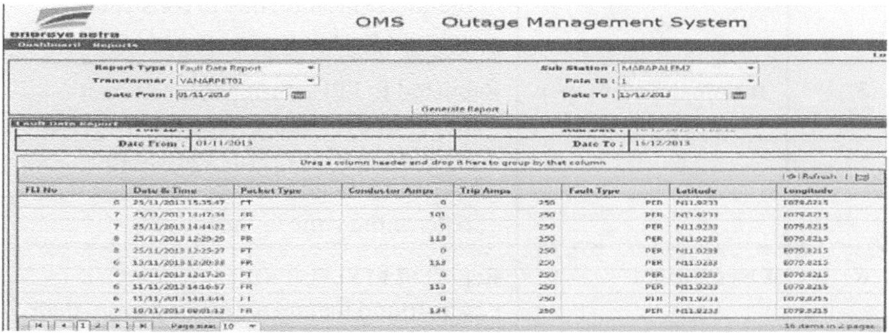

Figure 3.24 Fault data report of OMS.

of tampering include overcurrent, load through the earth, and opening of terminal cover. There are five actions reported under the opening of the terminal cover, two actions are reported for overcurrent, and one action reported for load through the earth.

Based on these tampering actions, thorough investigations were carried out by the Puducherry electricity department (PED). The reason for overcurrent was due to one customer exceeding the permissible current limit while that of load through the earth was due to the maloperation of the meter itself. The cause for the opening of the terminal cover was that three customers have shifted their locations of smart meters without informing the utility.

Table 3.6 Details and actions taken for tampering.

S. no.	Tamper detail	Action taken
1	Terminal Cover Open	Reported to PED, during inspection, it was found that the consumer shifted the meter location due to construction without the knowledge of PED staff.
2	Terminal Cover Open	Reported to PED, during inspection, it was found that the consumer shifted the meter location due to construction without the knowledge of PED staff.
3	Terminal Cover Open	Reported to PED, during inspection, it was found that the consumer bypassed the meter. PED warned the consumer that he will be penalized if he repeats the same.
4	Terminal Cover Open	Reported to PED, during inspection, it was found that the consumer shifted the meter location due to construction without the knowledge of PED staff.
5	Terminal Cover Open	Reported to PED, during inspection, it was found that the consumer connected welding machine. PED warned the consumer that he will be penalized if he repeats the same in the future.
6	Over current	Reported to PED, during inspection, it was found that meter reading more than actual consumption due to problem in the meter circuit and the same replaced with new meter.
7	Over current	Reported to PED, during inspection, it was found that due to a function at consumer house and overdrawing sanction load. PED warned consumer to get separate permission for over drawl.
8	Load through Earth	Reported to PED, during inspection, it was found that there is a damaged wiring inside the house. Instructed to consumer to renovate the wiring.

3.8   Case Study: Solar PV Awareness of Puducherry SG Pilot Project

The impact of solar energy awareness in the Puducherry SG pilot project is discussed in [5]. Puducherry is located in an area having low wind, and hence, wind generation is not possible. However, it has an annual average solar insolation level of 5.14 kWh/m²/day, and hence, it is suitable for solar power production. Hence, PGCIL in collaboration with PED has implemented a SG pilot project at Puducherry in 2014 under the prestigious "Smart Grid Vision Road Map" launched by the Union Ministry of Power in 2013. To study the impact of solar PV awareness in the Puducherry SG pilot project, a survey was taken from the 25 house owners living in the same area as that of the project.

This study focuses only on the rooftop solar PV installed in the Puducherry SG pilot project. This survey contains the data sample for 25 respondents. The demographic data for the customers is given in Table 3.7. Based on the income earned by the customers, they fall under different categories: 6 out of 25 respondents earn monthly income less than Rs. 15,000 or 15k. Nine out of 25 earn monthly income from Rs. 15,000 (or 15k) to Rs. 50,000 (or 50k). The remaining 6 earn monthly income greater than Rs. 50,000 (or 50k). Based on monthly energy consumption, 5 out of 25 consume from 0 to 250 units per month while 18 out of 25 consume from 251 to 800 units. Only 2 out of 25 customers consume greater than 800 units per month.

Table 3.7 Demographic data of the respondents.

Sr. no.	Description	Response	Response	Response
5	Monthly income in INR	(< 15k) 6	(15–50k) 9	(>50k) 6
6	Estimated monthly power consumption units in kWh	(0–250) 5	(251–800) 18	(>800) 2
8	Using star rated energy efficient appliances	(None) 2	(Some-most) 22	(All)

The house details of respondents are given in Table 3.8. 24 out of 25 respondents have their own house and the remaining 1 respondent lives in a rented house. Based on the type of house, 14 live in independent houses while the remaining 11 live in small/large apartments.

Table 3.8 House details of the respondents.

Sr. no.	Description	Response	Response	Response
16	House details	(Own) 24	(Rented) 1	(Leased) 0
17	House type	(Independent) 14	(Small/large apartment buildings) 11	(Govt./ society owned) 0
18	Availability of rooftop space	(Available) 22	(Partly/not available) 2	(Not shadow free) 1

The concerns/interest of the respondents is given in Table 3.9. The responses of the respondents are collected for various categories such as regularly used household appliances, types of electric bulb used, number of times the monthly electricity bill is reviewed, number of times power cut is experienced in a month, person/organization contacted when problem occurs with electricity service, respondents interested in energy saving and conservation measures, respondents interested in solar power generation other than rooftop, and person/organization benefitted from rooftop space (RTS) adoption.

Table 3.10 shows the awareness and attitude of respondents toward rooftop solar PV. The responses of the respondents for different queries such as willingness to use rooftop space in the house for solar power generation, source to find the solar power, the need for funding assistance provided by 70% bank loan, willingness to use 15% subsidy from Ministry of New and Renewable Energy (MNRE), willingness to use battery backup, and interest for rooftop solar PV generation is collected.

3.9   Recent Trends in Smart Grids

3.9.1   Smart GRIP Architecture

Adding new entries into the existing grid such as distributed energy sources, EMSs, and load controllers is a challenging task. The major concern is fluctuating renewable energy generation from the generation side and unpredictable energy demand from the customer side. The traditional control program can monitor and control up to a few thousand points. But with millions of connection points, this traditional control program faces a major challenge. Hence, a new architecture called the sensible grid with

Table 3.9 Concerns and interest of the respondents.

Sr. no.	Description	Response	Response	Response	Response
7	Regularly used household appliances	(Computer/TV/fridge) 24	(Kitchen mixer/grinder/geyser/water heater) 12	(Air conditioner/air cooler) 22	
9	Type of electric bulb used	(Incandescent) 2	(Fluorescent) 8	(CFL/LED) 25	
10	How often you review the monthly electricity bill?	(Never) 3	(Once/few) 17	(Very often) 5	
11	How many times in a month do you experience power cut?	(5–10) 19	(11–15) 1	(>15/Daily) 4	
12	When you have a problem with electricity service, you call	(PED) 22	(Neighbors/local resident association) 3	(MLA or political/prominent personalities) 2	
13	Interested in energy saving/conservation measures	(Yes) 25	(No) 0	(May be) 0	
14	Interested in Solar Power Generation other than Roof Top	(Yes) 21	(No) 2	(May be) 2	
15	Whom you think RTS adoption would benefit	(Consumer) 24	(MoP/MNRE/GoI) 18	(PED/REAP/GoP) 9	

Table 3.10 Awareness and attitude of the respondents toward rooftop solar PV.

Sr. no.	Description	Response	Response	Response
19	Would you like to use rooftop space of your house for solar power generation?	(Yes) 21	(No) 0	(May be) 4
20	How did you find out about solar power?	(REAP) 0	(PED) 1	(Other) 21
21	Do you require funding assistance of 70% through a bank loan	(Yes) 6	(No) 12	(May be) 3
22	Do you wish to avail 15% subsidy from MNRE	(Yes) 17	(No) 2	(May be) 3
23	Do you prefer to have a battery backup	(Yes) 8	(No) 12	(May be) 1
24	For rooftop solar, interested in	(Net metering) 14	(Feed in tariff with subsidy) 4	Net banking/adjustment 2

intelligent periphery, i.e., smart GRIP, is proposed [8] in response to the challenges faced by the conventional SG. The three pillars based on which smart GRIP is developed are empowering the periphery, abstracting commonality, and layered architecture. It is shown in the Figure 3.25.

1. Empowering the periphery: The periphery of the future grid comprises of micro-grid, distribution grid, industries, factories, EMSs, and houses. It includes uncertainties posed by both generation and customer. The major solution for this problem is to manage the uncertainty near to the source itself, i.e., the periphery.

2. Abstracting commonality: The future distribution systems comprises of micro-grid with bidirectional power flow and local generation sources. The core consists of the transmission system with real-time EMS while the periphery consists of micro-grids, distribution grids, homes, factories, and industries. The key approach for future control strategy is to explore the commonality between the core and periphery.

3. Layered architecture: Today's traditional grid is focused toward the core while tomorrow's future grid or Energy Internet is focused toward the periphery. The traditional

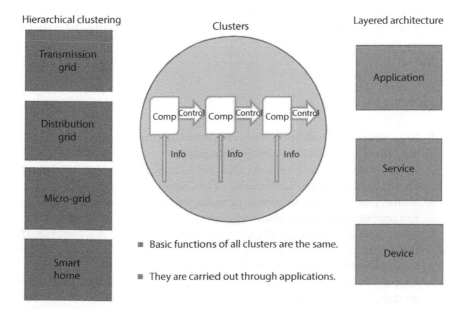

Figure 3.25 Smart GRIP architecture.

system has started to adopt new technologies at a slower rate. The evolution of today's legacy grid will happen at a varying rate. All the components are clustered with a control mechanism that permits the flow of both power and information.

3.9.2   Implementation of Smart Meter With Prepaid Facility

One of the latest innovations in smart meter is the development of prepaid smart meters [9], i.e., smart meters with prepaid facility as we have now for mobile phones and DTH. The customer had to recharge the smart meter by paying a considerable amount. There is a minimum balance amount that is used as a threshold value. The current balance amount is compared with this minimum balance. If the current balance amount is less than the minimum balance, an alert is given for the customer to recharge. When the current balance is greater than the minimum balance, the smart meter computes the total energy consumption and its corresponding tariff amount. The tariff amount is then subtracted from the current balance.

If the current balance becomes zero, then a warning message is sent to customer and a grace time of 30 minutes is given for the customer to recharge. If the customer fails to recharge after the grace time, then the power supply is switched off and an intimation message is sent to the customer. Once the customer receives this message, he/she can contact the utility or the service provider and receive supply after paying the penalty charges fixed by the utility.

References

1. India Smart Grid Roadmap, 2016, (https://indiasmartgrid.org/reports/India%20Smart%20Grid%20Roadmap_2016.pdf).
2. India Country Report on Smart Grids, 2017, (https://dst.gov.in/sites/default/files/India%20Country%20Report%20on%20Smart%20Grids.pdf).
3. Ministry of Power, Government of India website(https://powermin.nic.in/en/content/power-sector-glance-all-india).
4. *Smart Grid Handbook for Regulators and Policy Makers*, November 2017, (https://indiasmartgrid.org/reports/Smart%20Grid%20Handbook%20for%20Regulators%20and%20Policy%20Makers_20Dec.pdf).
5. Kappagantu, R. and Arul Daniel, S., Challenges and issues of smart grid implementation: A case of Indian scenario. *J. Electr. Syst. Inf. Technol.*, 5, 453–467, 2018.

6. Yang, B., Liu, S., Gaterell, M., Wang, Y., Smart metering and systems for low-energy households: challenges, issues and benefits. *Adv. Build. Energy Res.*, 13, 1, 80–100, 2019.

7. Kappagantu, R., Arul Daniel, S., Suresh, N.S., Techno-economic analysis of Smart Grid pilot project- Puducherry. *Resour.-Effic. Technol.*, 2, 185–198, 2016.

8. Felix, F., Varaiya, P.P., Hui, R.S.Y., Smart Grids with Intelligent Periphery: An Architecture for the Energy Internet. *Engineering*, 1, 4, 436–446, 2015.

9. Kathiresh, M. and Subahani, A.M., Smart Meter: A Key Component for Industry 4.0 in Power Sector, in: *Internet of Things for Industry 4.0. EAI/Springer Innovations in Communication and Computing*, Kanagachidambaresan, G., Anand, R., Balasubramanian, E., Mahima, V. (Eds.), 2020.

4

Internet of Things–Based Advanced Metering Infrastructure (AMI) for Smart Grids

V. Gomathy[1], V. Kavitha[2], C. Nayantara[3],
J. Mohammed Feros Khan[4], Vimalarani G.[5] and S. Sheeba Rani[1*]

*[1]Department of EEE, Sri Krishna College of Engineering and Technology,
Coimbatore, India*
[2]Dept. of CSE, Prathyusha Engineering College, Chennai, India
[3]Dept. of EEE, Sri Sairam Engineering College, Chennai, India
[4]Dept. of EEE, Vickram College of Engineering, Madurai, India
[5]Dept. of ECE, Hindustan Institute of Technology and Science, Chennai, India

Abstract

With the increase in global energy demand, the need for efficient metering infrastructure and energy wastage reduction is essential. Smart cities with smart grid (SG) technology can use smart meters for data collection and transmission based on advanced metering infrastructure (AMI). This helps in efficient analysis and storage of data and significant improvement in power management. A well-organized energy demand management system enables proper allocation of available resources. Precise analysis and forecasting the future energy needs enable optimal energy management. People and things are connected via Internet of Things (IoT) at any time and place. There is a significant contribution that can be made by IoT to SGs. The IoT-based AMI for SGs enables real-time energy management with reduced delay and product cost. It also enables remote billing and data visualization while providing significant data security.

Keywords: Advanced metering infrastructure, Internet of Things, smart grid, smart meter, cloud storage

*Corresponding author: sheebaranis@skcet.ac.in

M. Kathiresh A. Mahaboob Subahani and G.R. Kanagachidambaresan (eds.) Integration of Renewable Energy Sources with Smart Grid, (77–100) © 2021 Scrivener Publishing LLC

4.1   Introduction

The traditional electric grids used in the 1800s were replaced by power grids by certain countries in the 1960s. The economic and technical boom due to the centralized power generation of nuclear, hydro and fossil plants was remarkable during this period [1]. Sufficient quality, reliability, and high load delivery capacity were offered by the power distribution networks. During the later portion of the 20th century, with the increase in number of customers, the energy demand increased drastically as the ventilation, heating, cooling, and entertainment systems were significantly dependent of electricity [2].

A prominent fluctuation was experienced at the rate of power consumption. The power quality dropped with the increased peak hour power demands. In order to avoid voltage drops and the power fluctuations, the number of power generation plants had to be increased. However, the installation of new plants was expensive [3]. During the night time, the rate of power consumption was much lower that left the production capacity of the plant idle. This caused serious imbalance in the system. Demand side management (DSM) scheme was introduced by the electricity industry by offering several incentives for encouraging consumers toward management of energy utilization to promote an even consumption pattern.

The socioeconomic growth of any nation is greatly influenced by its energy demand and consumption [4]. In developing cities, urbanization leads to increase in commercial and residential load. This may lead to a gap in the demand supply ratio. It is impossible to imagine live without electricity in the current scenario. Energy crisis is a serious issue faced by several cities. This issue can be overcome by increasing the production of power. However, the depletion of fossil fuels and sources of power production poses the need to optimize the power consumption and rely of renewable energy sources [5]. However, energy production using renewable sources is expensive and intermittent. The other significant challenges involve requirement of improved control system for complex power models, economic upgradation in terms of network expansion and equipment renovation, system security from physical and cyber threats and consumer data privacy, environmental concerns, delivery power quality improvement, and the use of distributed energy resources (DERs). An optimal energy management system helps in overcoming most of these challenges.

4.1.1   Smart Grids

Energy crisis issue is faced by several developing countries in the recent years. When compared to the amount of energy produced, the population growth and industry development has been rapid leading to a huge energy demand [6]. The limited energy sources are another reason for the mismatch in the energy demand-supply ratio. Increasing the power generation and utilities to meet the growing consumer demand is expensive. The traditional equipment of the conventional electrical system does not satisfy the advanced technical necessities [7]. To maximize the efficiency of the power sources, it is essential to implement new technology in the conventional electrical system.

The smart grid (SG) infrastructure is introduced to overcome the challenges of the conventional energy system. It enables well-managed and continuous power delivery. SGs enable bidirectional communication. It makes use of smart meters for continuous monitoring and adjusting the consumption of electricity [8]. These help in enhancing the energy management model in home environment. During the low-demand period, the energy-intensive tasks can be scheduled by the consumers and utility. The expensive power plants that are built and maintained these days are used only when there is critical energy demand. During the peak periods, the power utilization can be reduced by direct interaction with the end-user equipment using SGs [9]. The digital information that can be instantly retrieved on the SGs provides scope for several improvements in the system. This helps in reducing the maintenance and operational costs while increasing the system efficiency.

The concept of SGs evolved in the 21st century with the advancement and innovation in various sectors [10]. Adaptive billing techniques were introduced with the development of smart sensors that overcame the need for accurate measurement of power utilization at the consumers end. These information and communication technology (ICT) innovations provided financial motivation that enabled consumers to shift power consumption during off-peak hours. Decentralization of electricity generation is enabled with the advancements in geothermal, tidal, solar, wind, and such renewable energy plants and technology integrated smart energy storage systems [11].

Several nations are switching to SGs with regard to modernizing their energy generation and distribution [12]. Electrical networks are combined with communication devices to provide an integrated energy system termed as SG. Real-time data can be exchanged in these energy efficient bidirectional systems. When compared to the conventional electric power system, SG can manage the generation and utilization of any type of power efficiently while providing interference resistance, real-time pricing of issues, self-healing, and improved power quality.

Figure 4.1 summarizes the progression of sub-systems with respect to SGs, their roles and relationship [13]. Data is gathered from the consumers, power generation plants and other network components with the use

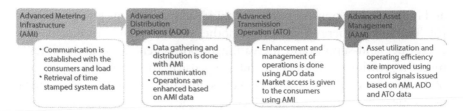

Figure 4.1 Overview of the SG sub-system progression.

of ICT to improve the reliability and system efficiency in SGs. This overcomes several drawbacks of the traditional electric grids [14]. The design of SGs involves several intelligent technologies that can be continuously enhanced and upgraded. However, interaction between the optimization, control and communication features of SG design must be unanimous to obtain maximum efficiency of the model. Prediction, adaptability and reliability issues of the traditional power grids can be addressed by an ideal SG [15]. Restoration, self-healing, performance, security, asset, and cost optimization, service and power quality, market enabling, end-to-end control capability, sustainability, flexibility, integration of advanced services, demand adjustment, and load handling are some of the various challenges that an SG should address [4].

4.1.2   Smart Meters

Reliable and cost efficient communication is provided in SGs using smart meters for reading the power consumption related information and transfer it to the service provider for invoicing purposes. Frequent monitoring and daily reporting functions can also be performed by smart metering system for gas, electric, and water utilities. The smart meter can transfer data to the service provider at predetermined equal time intervals enabling data gathering, work management, customer billing, data reduction, peak demand, distribution planning, distribution network analysis, quality of service monitoring, service restoration, service interruption, outage management, and time-based demand data analysis [16]. In case of fault occurrence in the conventional electric system, data is to be collected manually by the utility worker from the faulty location. However, real-time troubleshooting can be performed in SGs with the help of smart meter even without direct manual interaction.

Figure 4.2 provides the contrast between the conventional and smart energy meter. The manual data collection and billing process of the conventional system is completely automated in the smart system with

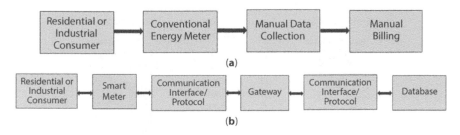

Figure 4.2 (a) Conventional and (b) smart energy meter.

two-way communication between the consumer and the database. There have been significant advancements in smart metering systems over the last few years. There are several specific communication standards available for the electric, gas, and water distribution systems [17]. This diversity is due to the variation in the kind of data that is obtained and transferred to the metering devices.

4.2   Advanced Metering Infrastructure

The smart meters have evolved from advanced meter reading systems to advanced metering infrastructures (AMIs). This enables prediction of power consumption and informed decision making by the consumers and demand-response management at the supply station [18]. The distribution networks for utilities like gas, electric and water make use of AMI. Monitoring and controlling the distribution for the required consumption levels need appropriate surveillance of the consumer demand. Communication between the utility servers and the meters is the major constituent of AMI. For this purpose, various communication technologies have been tested and instigated. Certain factors that are to be considered while installing the suitable communication infrastructure includes-taking the area of installation into consideration, communication permittivity with external devices, modification of network on installation of new meter, and training the metering maintenance personnel [19].

The collectors and smart meters are offered connectivity in terms of communication with the help of AMI as in Figure 4.3. Data is gathered from several devices and processed, and the results are reported over the communication network by the AMI. Advanced services like software updates, fault tolerance, intrusion detection, remote home control, billing, alarm, and monitoring can be provided by the AMI. The grid stability can be maintained by suitable monitoring of the SG [20]. Failure of

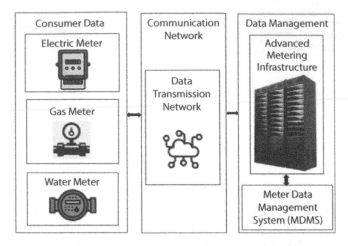

Figure 4.3 Basic advanced metering infrastructure model.

components in the grid or smart home alarms can be addressed by the alarm functionality. Maximum usage or hourly energy consumption can be estimated under the billing functionality. Interaction with the home appliances can be used for controlling the devices using remote home control functionality. Any attempts of hacking into the home can be detected and alerted using the intrusion detection functionality.

Unlike SG, AMI integrates several technologies into its configured infrastructure to achieve its goals. The AMI infrastructure is inclusive of data collection and data management system that comprises of smart meters at the consumers end and software application platform at the supplier end. Time-based information is gathered by the solid state electronic meter equipped at the consumer end. The data gathered by this meter is transmitted using the fixed networks that are available [21]. The AMI host system receives the metered information. This data is further transferred to the data management system for storage and analysis. Further, information is provided to the utility service provider in useful format. The two-way communication of AMI enables issuing of signals by the utility provider regarding price to the load controlling device or meter.

4.2.1   Smart Devices

Advanced electronic software and hardware comprising the end user devices enable measurement and data collection with regard to power utilization at required time intervals with time stamping feature. The system administrator sets the time intervals at which data is to be transmitted to the remote data center [22]. Bidirectional communication feature of AMI enables the load controlling devices and smart devices to receive command signals and take appropriate actions. Data is communicated between the service provider as well as the consumer through smart meters. The consumers can get information regarding the energy usage from the smart devices through the in-home display unit. The pre-defined directives and user criteria is used for regulation of power consumption in coordination with the pricing data provided by the service provider [23].

Thermal, fluid, and electrical smart meters can be used based on the application. The utility consumption is also influenced by light, temperature, humidity, and other factors which are measured by various sensors in the system. Depending on the functionality, cost, and requirements of the system designer or user, the sensors used in the system can be increased. Measurement of data and communication are the two major functions of the smart meter. For this purpose, metrology and communication sub-systems are enabled in every meter [24]. Factors like application, data security level, technical requirement, accuracy, measured values, and region influence the metrology of the system.

The communication scheme is decided based on factors like encryption and security. Irrespective of the measurement quantity or type, the meters should have certain basic functionalities [25]. Some of them include the following: time synchronization for reliable data transfer to a central hub for billing or data analysis; power management to enable the system to sustain its functionality; display for providing data to the consumers regarding consumption, billing, and demand management; communication channel for transmission and reception of operational commands and firmware upgradation; calibration and control of the meter to offset minor system variations; and accurate data measurement using various methods, topologies, and physical principles [26].

The significant features of smart electricity meters include efficient power consumption, emission reduction, improvement of environmental conditions, interaction with smart devices, detection of energy theft, monitoring power quality in terms of power factor, active and reactive power, current, voltage and phase, demand response load limiting, remote control and command operations, outage and failure notification, net metering, utility and consumer consumption data display, and time-based pricing [27].

4.2.2 Communication

The data that is gathered must be transmitted by the smart meters for analysis to a computer. It should also be able to receive the instructions from the operation facility. The typical two-way communication is a vital feature of the AMI. The voluminous data needs to be transmitted through a highly reliable communication network due to the large number of consumers and smart meters at every terminal [28]. Ability for future expansion and technological upgradation, precision of communication, data authenticity, cost effectiveness, grid status display, sensitive data confidentiality, data access restriction, and voluminous data transfer are some of the key factors that are to be considered while designing and selecting a suitable communication network.

Several topologies are available for SG communication. The commonly used architecture uses local data concentrators for gathering information from groups of meters and transfer the information to a central command using a backhaul channel. This consists of billing applications, data processing, storing and management facilities, as well as the servers. Several communication mediums and technologies are available to match the diverse networks and architectures exist for implementation of AMI. Some of them are Zigbee, peer-to-peer, GPRS, Bluetooth, WiMax, cellular, optical fiber, copper wire, Broadband over power lines (BPL), Power Line Carrier (PLC), and so on. Utility networks enable communication between the utility provider and the Home Area Networks (HAN), while the in-home network allow communication between the household devices in the AMI level [29].

The controllers, in-home display (IHD), electric vehicles, energy generation, and storage such as wind and solar, smart devices in the home and surroundings, as well as the smart meters are connected by the HANs. Based on the task, the bandwidth of HAN varies for each device between 10 and 100 kbps due to the instantaneous data flow. When large houses or office buildings are to be covered, the data rate and number of devices may increase. This requires an expandable network. The usage and loads are not critical which forms the basis for calculation of accepted delay and reliability. Different manufacturers produce smart meters with different features, benefits, and disadvantages. A single model with all the necessary technical features to suit all the requirements of a real-time environment is not available. The interoperability between smart meters from different manufacturers is impossible due to the lack of a single communication channel. Manufacture independent Open Meter project is proposed by OMS Group to overcome this drawback [19].

The available communication technologies can be categorized into five major zones for detailed protocol analysis. Remote as well as local communication is considered for offering improved flexibility. Figure 4.4 consists of direct communication, meter access network, meter gateway network, telecommunication service provider network, and cellular technology.

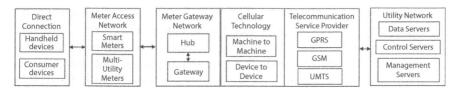

Figure 4.4 Various communication technologies for AMI.

Each zone performs a distinct function. Device-to-device or machine-to-machine communication is performed using the cellular technology through internet between the utility center and smart meter. The utility center is connected to the gateway through internet using telecom service provider network. Hubs are used to increase scalability using meter-gateway network [16]. Meter access network enables communication between smart meters in the same location. Smart meters are connected to the customers' devices or handheld units via direct connection.

Wireless fixed network, wireless mobile, wired, or a blend of them can be used in the communication technologies for data gathering and transporting. Alarms, message frequency, and data are included in the model. Economical aspect knowledge maturity, operation scheme within the utility, advantages of information utilization, deployment field configuration, and challenges faced by the utilities are certain deciding factors for selection of communication technology. Mobile satellitecommunication, Long Term Evolution (LTE), IEEE 802.16, 3G inclusive of HSPA+, HSPA, High Speed Downlink PacketAccess, UMTS, Telecommunication System and Universal mobile, 2G Technologies including EGPRS, Enhanced GPRS, GPRS (General Packet Radio Service), and GSM (Global System for Mobile Communication), ReFLEX, ERMES, and Terrestrial Trunked Radio (TETRA) can be of great use in the wireless applications. Wired technologies include Fiber to the Home or Fiber to the Building (FTTx), Asymmetric Digital Subscriber Line (ADSL), and Public Switched Telephone Network (PSTN) [28].

4.2.3   Data Management System

Billing requirements can be addressed by a system that can gather and analyze information at the utility provider end. The power consumption profile, demand response, and responses in the real-time environment for grid variations and emergencies should be handled by this system. A multi-modular structure consists of Transformer Load Management (TLM), Geographic Information System (GIS), Mobile Workforce Management (MWM),

Enterprise Resource Planning (ERP), load forecasting and power quality management systems, Outage Management System (OMS), utility website, billing system, Consumer Information System (CIS), and the Meter Data Management System (MDMS). The essential analytical tools for communication with the corresponding modules are integrated within the MDMS that acts as the central module of the management system [19].

The lower layer interruptions must be overcome to enable information flow to the management modules from the customer in an accurate manner and the AMI data must undergo Validation, Editing, and Estimation (VEE) in this system. Voluminous data is gathered by the AMI at regular time intervals. This data is termed as Big Data. Special tools are required for analysis and management of big data. The distribution network automation system that gathered real-time data at the rate of 30 samples per sensor per second, the smart meters (AMI), third-party systems that are linked to the grid such as DERs or storage, and asset management system that enables communication between the smart elements in the network and the central command for applications like firmware upgradationcontribute to the big data in SGs.

The design, applications, and features of MDMS vary from vendor to vendor. While certain systems provide data for other applications to use, some offer inbuilt application suites. Customer engagement, utility management and utility grid optimization, and enhancement are the major demands that are to be addressed by all MDMS irrespective of their complexity and features [14]. Data analytics is gaining significant attention in SG. Data mining techniques are used to gather all available internal and external data from the grid and are analyzed to make decisions based on valuable data that is extracted. The system needs the following components from the hardware and infrastructure perspective—a building to host the system termed as the data center infrastructure, ventilation, backup power and other auxiliary systems, data handling hardware and servers, storage system, data virtualization and analysis systems, and corresponding database software for resourceful utilization of computing resources and discrete storage.

Utility providers offer great attention to data storage due to the significance of the gathered information. Meticulous planning and design of storage facilities has to be done to make sure they are disaster proof and provide necessary contingency plans and back up under diverse scenarios. Implementation of these provisions are expensive. This can be overcome by using cloud computing and virtualization solutions. The investment return and efficiency can be improved by merging all resources through virtualization [20]. However, the complexity of the system increases with the additional technology requirements. Virtual resources from multiple locations can be accessed using cloud computing. However, the data security may be compromised using this technology. Different locations have different laws and regulations that relate to the gathered data, which can impose issues while implementing cloud computing. Different service providers and their capacity is used in cloud computing, thereby reducing the need and price of special purpose data centers.

4.2.4   Mathematical Modeling

Runtime power limit (PL_R) and maximum power limit (PL_{MAX}) can be used for characterization and modeling the smart energy meter.

$$PL_R = PL_{APL} \times PL_{MAX} \tag{4.1}$$

where PL_{APL} is the allowable percentage of received power limit. Agricultural, residential, commercial, and industrial power which are the major load consumption categories can be modeled mathematically for efficient power management. Agricultural power is represented by P_A; Residential power is represented by P_R; Commercial power is represented by P_C; Industrial power is represented by P_I; α, β, γ, and δ represents the power factor for agricultural, residential, commercial, and industrial power, respectively.

Agricultural consumer power allotment

$$P_{A(total)} = \alpha P_{A1} + \alpha P_{A2} + \alpha P_{A3} + \cdots + \alpha P_{An} \tag{4.2}$$

$$P_{A(max)} = \alpha \sum_{i=1}^{n} P_{An} \tag{4.3}$$

Residential consumer power allotment

$$P_{R(total)} = \beta P_{R1} + \beta P_{R2} + \beta P_{R3} + \cdots + \alpha P_{Rn} \tag{4.4}$$

$$P_{R(max)} = \beta \sum_{i=1}^{n} P_{Rn} \tag{4.5}$$

Commercial consumer power allotment

$$P_{C(total)} = \gamma P_{C1} + \gamma P_{C2} + \gamma P_{C3} + \cdots + \gamma P_{Cn} \tag{4.6}$$

$$P_{C(max)} = \gamma \sum_{i=1}^{n} P_{Cn} \tag{4.7}$$

Industrial consumer power allotment

$$P_{I(total)} = \delta P_{I1} + \delta P_{I2} + \delta P_{I3} + \cdots + \delta P_{In} \tag{4.8}$$

$$P_{I(max)} = \delta \sum_{i=1}^{n} P_{In} \tag{4.9}$$

Total power allotment

$$P_{max} = P_{A(max)} + P_{R(max)} + P_{C(max)} + P_{I(max)} \tag{4.10}$$

Agricultural consumer power demand

$$P_{A(td)} = \alpha P_{A1d} + \alpha P_{A2d} + \alpha P_{A3d} + \cdots + \alpha P_{And} \tag{4.11}$$

$$P_{A(td)} = \alpha \sum_{i=1}^{n} P_{And} \tag{4.12}$$

Residential consumer power demand

$$P_{R(td)} = \beta P_{R1d} + \beta P_{R2d} + \beta P_{R3d} + \cdots + \alpha P_{Rnd} \tag{4.13}$$

$$P_{R(td)} = \beta \sum_{i=1}^{n} P_{Rnd} \tag{4.14}$$

Commercial consumer power demand

$$P_{C(td)} = \gamma P_{C1d} + \gamma P_{C2d} + \gamma P_{C3d} + \cdots + \gamma P_{Cnd} \tag{4.15}$$

$$P_{Ctd} = \gamma \sum_{i=1}^{n} P_{Cnd} \tag{4.16}$$

Industrial consumer power demand

$$P_{I(td)} = \delta P_{I1d} + \delta P_{I2d} + \delta P_{I3d} + \cdots + \delta P_{Ind} \tag{4.17}$$

$$P_{I(td)} = \delta \sum_{i=1}^{n} P_{Ind} \tag{4.18}$$

Total power demand

$$P_{td} = P_{A(td)} + P_{R(td)} + P_{C(td)} + P_{I(td)} \tag{4.19}$$

The total available power, allotted power, and demand can be compared and energy requirement can be managed. The maximum permitted limit is set along with a flexible value range for each consumer. This data is transmitted over the Internet of Things (IoT) to the smart energy meter.

4.2.5 Energy Theft Detection Techniques

Game theory–based detection, state-based detection, and classification-based detection schemes are used for identifying energy theft. The categorization is as represented in Figure 4.5.

4.3 IoT-Based Advanced Metering Infrastructure

This system enables forecasting utilization of energy in the future and set a trigger level where an alert is sent if the consumption exceeds the predefined value. The significant features of this model includes turning the appliances on/off automatically when not in use based on the occupancy detection algorithm and to facilitate consumer participation in energy management by measuring bidirectional energy flow [23]. Peak hour load shedding can be performed by categorizing loads into various levels. This enables overcoming power quality issues and peak load shaving by the system operator using IoT for remote access to control the appliances enabling efficient energy management. The next-generation smart building management and energy optimization schemes are greatly influenced by IoT. Certain communication protocols are used for monitoring and remote control in the system. The load requirements of each building can be categorized as heavy, medium and low loads. Figure 4.6 represents the AMI.

Based on the load type, the information regarding the measured electrical parameters is provided by the smart meter. On occurrence of power

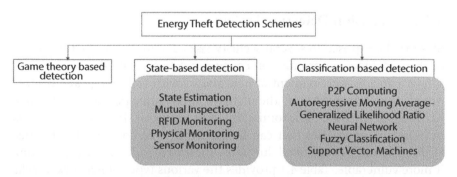

Figure 4.5 Energy theft detection schemes—classification.

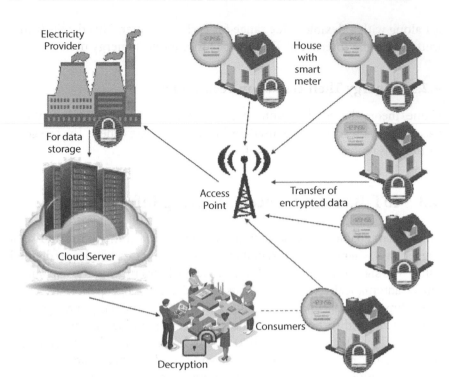

Figure 4.6 Advanced metering infrastructure.

quality issues and emergencies, suitable actions can be taken by the operator on the basis of the information provided by the data log through data visualization [26]. The number of occupants in a room can be predicted accurately with smart means like occupancy prediction algorithm. The system operator can use the IoT platform for remote control of the gadgets and appliances.

4.3.1   Intrusion Detection System

SGs may be exposed to several security risks due to various factors. Some of them include compromise on consumer data security due to data volume expansion, development of new attacks with technological advancements, attacks facilitated by the increase in the entry points in to the SG, hardware or software compromise, Internet Protocol (IP) communications network susceptibilities that can be caused by interconnected networks, and attacks can be facilitated due to the increased grid complexity, making it more vulnerable. Table 4.1 provides the various types of network attacks on SGs, the targeted medium, and activities of the attacker on the target.

Table 4.1 Types of smart grid network attacks.

Attack type	Target	Attacker activities
Compromised key	Communication Network	Access to user progress is secured by procuring the key.
Identity spoofing	Computer Network	User system is accessed by source IP falsification.
Application layered	Application Network	Errors are caused deliberately on a targeted application.
SQL injection	Computer Network	User functionality is disrupted by an SQL query injection.
Denial of service	Computer Network	User access is prevented.
Man in the middle	Communication Network	User actions are monitored by intrusion as a middle-man.

It is essential to defend the system from these attacks while keeping the SG network unharmed and protecting the data security of the consumers.

Network attacks are often defended using cryptographic techniques and hash functions. Data is scrambled electronically in cryptography scheme. It offers data authorization, authentication, integrity, and confidentiality. Confidentiality during data storage and transmission is ensured by encryption and decryption schemes. Data is converted to a code using encryption and has to be decrypted to be readable. Data cannot be altered without the required permission, thereby ensuring its integrity. Processing and moving of data is performed only in predictable manner. Data is converted into strings of fixed size using cryptographic hash function.

Intrusion-Based Detection System (IDS) is commonly used in SG in the recent days as a defense mechanisms against network attacks. The network is monitored continuously and benign or malicious intrusions are distinguished. The knowledge from previously known attacks are used for analyzing the attack pattern, examination of file integrity and system vulnerability by monitoring the system activity. Future threats are identified by searching the internet and automatic monitoring [15]. If any security policy violation, malicious activity, or compromise of system information occurs, the administrators are alerted automatically by the IDS security software. Intrusion of unauthorized entities are identified by continuous monitoring of the network traffic. However, sometimes, the attacks are identified after the network is compromised and damaged.

Normal activities as well as suspicious or forbidden activities are detected by the IDS. For this purpose, the system has to understand the difference between these activities. Artificial intelligence can be used for training the system to differentiate between these activities. Intrusion detection techniques and algorithms based on genetic algorithms, neural networks, probability reasoning, or fuzzy logic can be used for this purpose [30]. Negative Selection Algorithm (NSA) based on Artificial Immune System (AIS) scheme is also used for defending SG against attacks [31]. This model determines malicious patterns and stop the attacks before they take place.

Figure 4.7 provides the hardware structure of the IoT-based AMI system. Both wired and wireless communication schemes can be used for data transfer. The occupancy is detected using IoT-based wireless proximity sensor and IR sensor and the data is analyzed using an Arduino Uno processor and the load is turned on or off based on the instructions from the program. The load and power consumption–based information can be displayed in the display unit or user devices over the IoT platform.

The various techniques that enable location and estimation of power theft may work independently or along with the smart meters. Harmonic Generator, Power Line Impedance, Support Vector Machine [32], Genetic Algorithm, and Central Observer Meter are some of the location schemes. The mathematical approaches used for this purpose include Optimum Path Forest classifier (OPF) [33], Artificial Neural Network Multi-Layer Perceptron (ANN-MLP), Support Vector Machine Radial Basis Function (SVM-RBF) [34], and Support Vector Machine Linear (SVM-LINEAR) [35].

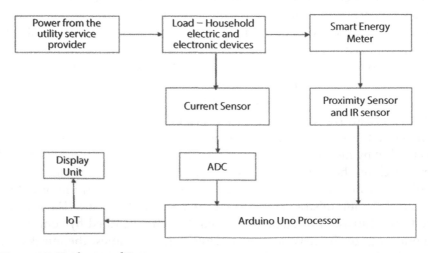

Figure 4.7 Hardware architecture.

4.4   Results

The information from smart meters can be used for creating consumption patterns for every household or commercial building. Based on the geographic area, industry or user, the consumption patterns vary. Gas, water, or electricity management studies can be carried out by the utilities with these patterns. Simulation of the IoT-based AMI is performed on LabView platform. Energy and power is analyzed. The filter circuit is fed with the input voltage and current. Noise from the waveform is removed. The filtered wave undergoes power spectrum and fundamental vector analysis after calculation of energy and power. Based on the load consumption, sampling of output is done.

The power demand can be predicted. This can be used for adjusting the power consumption of the utilities. User profiles can be created using these patterns with regard to the amount of usage of certain devices, time spent at home, and so on. Along with this, the data from smart meters can be used for detection of gas or water leakage. The signaling, one-cycle power and total power measurement can be simulated. The values from current and voltage spectra can be used for calculation of power. Power factor, reactive power, active power, and apparent power analysis can be performed. The consumer behavior like habitability and peak consumption can be predicted using this model.

Table 4.2 provides the load capacity of some test cases that were implemented during simulation. Different combination of these loads is tested to analyze the corresponding power, current and voltage. The actual power consumption is compared to the displayed values and the error is analyzed.

Table 4.3 provides various test cases for different loadcombinations. The actual current and power along with the display current and voltage and the error is analyzed in Table 4.4 with a standard voltage input for all the

Table 4.2 Load capacity.

Load	Power rating
Lamp	18 Watts
Fan	50 Watts
Refrigerator	150 Watts
Computer	200 Watts
Air Conditioner	500 Watts

Table 4.3 Load types.

Case	Combination
1	Lamp + Fan
2	Lamp + AC
3	Lamp + Fan + Refrigerator
4	Lamp + AC + Refrigerator
5	Lamp + Fan + Computer
6	Lamp + Fan + Computer + Refrigerator
7	Lamp + Computer + AC
8	Lamp + AC + Computer + Refrigerator

test cases. Figure 4.8a represents the load curve for the various test cases over a stretch of 24 hours. The load profile is represented in Figure 4.8b. Separate information is provided regarding the different load types. Based on the energy consumption time peak or general hours, the price per unit is fixed. The total hours of operation and cost of power consumption will be displayed at the display unit.

The power consumption on weekends and holidays is represented in Figures 4.9a and b represent the annual power consumption analysis. The least power consumption period is identified and assumed that the consumer is not available in the building during the period. These cases can be applied to rural houses, urban houses, and businesses as the energy consumption pattern accuracy and efficiency of data management is high. This enables improving the billing and settlement audit process, demand profiling, customer segmentation, localized servicing, distribution planning, automated prediction and action, system monitoring, launching tariffs and energy reduction programs, theft and tamper detection, reduction of unbilled energy, data collection cost reduction, as well as asset failure reduction.

4.5 Discussion

Smart meters and AMI offers several benefits. Manual meter reading is avoided and hence staff saving can be done. The gas, water, and power supply quality and reliability are increased. Easy problem identification,

Table 4.4 Load analysis.

Case	Voltage	Actual current	Display current	Error	Actual power	Display power	Error
1	230 V	0.29	0.28	0.01	68	70	2
2	230 V	2.25	2.23	0.02	518	520	2
3	230 V	0.95	0.94	0.01	218	217	−1
4	230 V	2.90	2.89	0.01	668	665	−3
5	230 V	1.17	1.15	0.02	268	270	2
6	230 V	1.82	1.81	0.01	418	412	−5
7	230 V	3.12	3.11	0.01	718	721	3
8	230 V	3.77	3.75	0.02	868	869	1

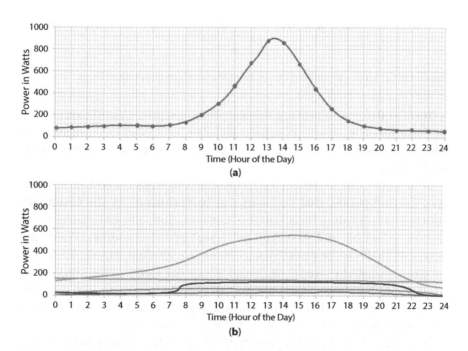

Figure 4.8 (a) Overall load curve. (b) Load capacity–based profile for each device.

analysis, and isolation are performed. Leak detection and repair are performed in a timely manner improving efficiency of consumption. Wastage of gas, water, and power is reduced, thereby maintaining environmental sustainability. The future power consumption forecasting can be

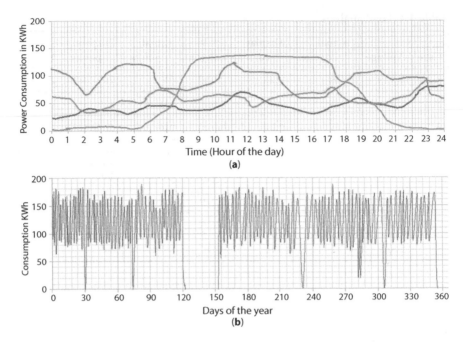

Figure 4.9 (a) Weekend and holiday. (b) Annual power consumption analysis.

performed thereby improving the distribution efficiency and overcoming the need for new power generation facilities by efficient management of peak power demand. Tampering and counterfeiting the network or data can be detected.

Billing cost can be reduced by efficient consumption and leak detection. Consumption-based decisions can be performed by the customers. Information regarding the amount of consumption, efficient management for cost reduction, and so on can be viewed by the consumer. Water and electricity theft can be detected. Customers are affected monetarily due to the gas, water, and power theft. It also affects the safety of the consumers, utilities, and buildings. Smart meters can provide information to the utilities for theft and fraud detection. Customer profiles and their corresponding consumption patterns can be compared for this purpose.

Tracking personal consumption habits help in detection of anomaly and identification of fraud. Data analysis is done such that the customer privacy is maintained. The confidentiality, integrity, availability, and accountability of data are preserved while protecting the data from physical and external cyber-attacks. Some of the major challenges faced while implementing AMI for SGs using IoT include privacy issues. Energy thieves are constantly developing new schemes to intrude into the security schemes.

Secure data gathering, efficient data processing, and storage costs have also imposed several challenges due to the volume of the data that is gathered.

4.6   Conclusion and Future Scope

IoT-based AMI is developed for smart grids. Utilities and smart meters are connected to the internet and communication protocols are used for information transfer and exchange. Quick and efficient energy management on SGs is performed with lesser product cost and higher data security. Voltage, current, and power factors are monitored continuously and data logs are maintained in the server. Remote monitoring of consumption data is enabled through IoT while data security is maintained. Future work is focused toward development of robust IoT protocols for voluminous data management and complete data loss reduction. Incorporation of cloud computing and big data techniques in a detailed manner, enabling trusted computing feature and enhancing privacy preservation, are other future directions of this work.

References

1. Mohammadali, A., Haghighi, M.S., Tadayon, M.H., Mohammadi-Nodooshan, A., A novel identity-based key establishment method for advanced metering infrastructure in smart grid. *IEEE Trans. Smart Grid*, 9, 4, 2834–2842, 2016.
2. Ghosal, A. and Conti, M., Key management systems for smart grid advanced metering infrastructure: A survey. *IEEE Commun. Surv. Tutor.*, 21, 3, 2831–2848, 2019.
3. Kakakhel, S.R.U., Kondoro, A., Westerlund, T., Dhaou, I.B., Plosila, J., Enhancing Smart Grids via Advanced Metering Infrastructure and Fog Computing Fusion, in: *2020 IEEE 6th World Forum on Internet of Things (WF-IoT)*, IEEE, pp. 1–6, 2020, June.
4. Albayati, A., Abdullah, N.F., Abu-Samah, A., Mutlag, A.H., Nordin, R., A Serverless Advanced Metering Infrastructure Based on Fog-Edge Computing for a Smart Grid: A Comparison Study for Energy Sector in Iraq. *Energies*, 13, 20, 5460, 2020.
5. Ribeiro, I.C., Albuquerque, C., Antônio, A.D.A., Passos, D., THOR: A framework to build an advanced metering infrastructure resilient to DAP failures in smart grids. *Future Gener. Comput. Syst.*, 99, 11–26, 2019.
6. Ghasempour, A. and Moon, T.K., Optimizing the number of collectors in machine-to-machine advanced metering infrastructure architecture

for internet of things-based smart grid, in: *2016 IEEE Green Technologies Conference (GreenTech)*, IEEE, pp. 51–55, 2016, April.

7. Chren, S., Rossi, B., Pitner, T., Smart grids deployments within EU projects: The role of smart meters, in: *2016 Smart cities symposium Prague (SCSP)*, IEEE, pp. 1–5, 2016, May.

8. CR, P.G., Ramesh, A., Satvik, D., Nagasundari, S., Honnavalli, P.B., Simulation of SCADA System for Advanced Metering Infrastructure in Smart Grid, in: *2020 International Conference on Smart Electronics and Communication (ICOSEC)*, IEEE, pp. 1071–1077, 2020, September.

9. Han, W. and Xiao, Y., Non-technical loss fraud in advanced metering infrastructure in smart grid, in: *International Conference on Cloud Computing and Security*, Springer, Cham, pp. 163–172, 2016, July.

10. Saghezchi, F.B., Mantas, G., Ribeiro, J., Al-Rawi, M., Mumtaz, S., Rodriguez, J., Towards a secure network architecture for smart grids in 5G era, in: *2017 13th International Wireless Communications and Mobile Computing Conference (IWCMC)*, IEEE, pp. 121–126, 2017, June.

11. Kamal, M. and Tariq, M., Light-weight security and blockchain based provenance for advanced metering infrastructure. *IEEE Access*, 7, 87345–87356, 2019.

12. Mahmood, K., Arshad, J., Chaudhry, S.A., Kumari, S., An enhanced anonymous identity-based key agreement protocol for smart grid advanced metering infrastructure. *Int. J. Commun. Syst.*, 32, 16, e4137, 2019.

13. Wan, L., Zhang, Z., Wang, J., Demonstrability of Narrowband Internet of Things technology in advanced metering infrastructure. *EURASIP J. Wirel. Commun. Netw.*, 2019, 1, 2, 2019.

14. Baskaran, H., Al-Ghaili, A.M., Ibrahim, Z.A., Rahim, F.A., Muthaiyah, S., Kasim, H., Data falsification attacks in advanced metering infrastructure. *Bull. Electr. Eng. Inform.*, 10, 1, 412–418, 2020.

15. Kabalci, Y., Kabalci, E., Padmanaban, S., Holm-Nielsen, J.B., Blaabjerg, F., Internet of Things applications as energy internet in Smart Grids and Smart Environments. *Electronics*, 8, 9, 972, 2019.

16. Ramakrishnan, R. and Gaur, L., Smart electricity distribution in residential areas: Internet of Things (IoT) based advanced metering infrastructure and cloud analytics, in: *2016 International Conference on Internet of Things and Applications (IoTA)*, 2016, January, IEEE, pp. 46–51.

17. Boudko, S., Aursand, P., Abie, H., Evolutionary Game for Confidentiality in IoT-enabled Smart Grids, Information, 11, 582, 2020.

18. Chen, S., Wen, H., Wu, J., Lei, W., Hou, W., Liu, W., Jiang, Y., Internet of Things based smart grids supported by intelligent edge computing. *IEEE Access*, 7, 74089–74102, 2019.

19. Jiang, R., Lu, R., Wang, Y., Luo, J., Shen, C., Shen, X., Energy-theft detection issues for advanced metering infrastructure in smart grid. *Tsinghua Sci. Technol.*, 19, 2, 105–120, 2014.

20. Vijayanand, R., Devaraj, D., Kannapiran, B., Support vector machine based intrusion detection system with reduced input features for advanced metering infrastructure of smart grid, in: *2017 4th International Conference on Advanced Computing and Communication Systems (ICACCS)*, IEEE, pp. 1–7, 2017, January.

21. Ghasempour, A., Optimized advanced metering infrastructure architecture of smart grid based on total cost, energy, and delay, in: *2016 IEEE Power & Energy Society Innovative Smart Grid Technologies Conference (ISGT)*, IEEE, pp. 1–6, 2016, September.

22. Ghasempour, A., Optimum number of aggregators based on power consumption, cost, and network lifetime in advanced metering infrastructure architecture for Smart Grid Internet of Things, in: *2016 13th IEEE Annual Consumer Communications & Networking Conference (CCNC)*, IEEE, pp. 295–296, 2016, January.

23. Singh, S.K., Bose, R., Joshi, A., Energy theft detection in advanced metering infrastructure, in: *2018 IEEE 4th World Forum on Internet of Things (WF-IoT)*, IEEE, pp. 529–534, 2018, February.

24. Li, H., Gong, S., Lai, L., Han, Z., Qiu, R.C., Yang, D., Efficient and secure wireless communications for advanced metering infrastructure in smart grids. *IEEE Trans. Smart Grid*, 3, 3, 1540–1551, 2012.

25. Selvam, C., Srinivas, K., Ayyappan, G.S., Sarma, M.V., Advanced metering infrastructure for smart grid applications, in: *2012 International Conference on Recent Trends in Information Technology*, IEEE, pp. 145–150, 2012, April.

26. Tudor, V., Almgren, M., Papatriantafilou, M., The influence of dataset characteristics on privacy preserving methods in the advanced metering infrastructure. *Comput. Secur.*, 76, 178–196, 2018.

27. Yan, Y., Hu, R.Q., Das, S.K., Sharif, H., Qian, Y., An efficient security protocol for advanced metering infrastructure in smart grid. *IEEE Network*, 27, 4, 64–71, 2013.

28. Xu, S., Qian, Y., Hu, R.Q., A study on communication network reliability for advanced metering infrastructure in smart grid, in: *2017 IEEE 15th Intl Conf on Dependable, Autonomic and Secure Computing, 15th Intl Conf on Pervasive Intelligence and Computing, 3rd Intl Conf on Big Data Intelligence and Computing and Cyber Science and Technology Congress (DASC/PiCom/DataCom/CyberSciTech)*, IEEE, pp. 127–132, 2017, November.

29. Román, L.F., Gondim, P.R., Lopes, A.P., A Lightweight Authentication Protocol for Advanced Metering Infrastructure in Smart Grid, in: *Anais do XII SimpósioBrasileiro de ComputaçãoUbíqua e Pervasiva*, 2020, June, SBC, pp. 21–30.

30. Mohassel, R.R., Fung, A.S., Mohammadi, F., Raahemifar, K., A survey on advanced metering infrastructure and its application in smart grids, in: *2014 IEEE 27th Canadian Conference on Electrical and Computer Engineering (CCECE)*, IEEE, pp. 1–8, 2014, May.

31. Yip, S.C., Wong, K., Hew, W.P., Gan, M.T., Phan, R.C.W., Tan, S.W., Detection of energy theft and defective smart meters in smart grids using linear regression. *Int. J. Electr. Power Energy Syst.*, 91, 230–240, 2017.
32. Shein, R., Security measures for advanced metering infrastructure components, in: *2010 Asia-Pacific Power and Energy Engineering Conference*, 2010, March, IEEE, pp. 1–3.
33. Parra, G.D.L.T., Rad, P., Choo, K.K.R., Implementation of deep packet inspection in smart grids and industrial Internet of Things: Challenges and opportunities. *J. Netw. Comput. Appl.*, 135, 32–46, 2019.
34. Bai, X.M., Jun-xia, M., Zhu, N.H., Functional analysis of advanced metering infrastructure in smart grid, in: *2010 International Conference on Power System Technology*, IEEE, pp. 1–4, 2010, October.
35. Amin, S., Schwartz, G.A., Cárdenas, A.A., Sastry, S.S., Game-theoretic models of electricity theft detection in smart utility networks: Providing new capabilities with advanced metering infrastructure. *IEEE Control Syst. Mag.*, 35, 1, 66–81, 2015.

5

Requirements for Integrating Renewables With Smart Grid

Indrajit Sarkar

Department of EE, NIT Rourkela, Odisha, India

Abstract

In the search for sustainable clean and green energy sources, integrating renewables with the conventional utility grid has become a standard practice. However, the power delivered by the renewable energy resources like wind and solar is intermittent in nature compared to conventional fossil fuel-based energy generations. For example, it is not easy to predict the availability of wind energy much in advance; on the other hand, even though the variability in solar energy is more than wind energy, its availability prediction is easier. Therefore, integrating such uncertain and variable energy sources with the utility grid involves appropriate planning and controlled operation of the power system as well as has extensive repercussions on the existing power transmission and distribution systems. The effective grid integration of renewable energy resources is the practice of developing efficient ways to deliver the variable renewable energy to the grid with appropriate power and control strategies by maximizing the cost-effectiveness and improved system stability and reliability. Therefore, it involves the policymakers, regulators, and system operators to consider a number of factors such as how to assimilate the new renewable sources, new transmission and distribution system, improving system flexibility, and planning for applicability of renewable resources in future. In this chapter, a brief introduction to the renewable energy resources and their characteristics is presented, and then, it is explained how the intermittent characteristics of the renewable resources affects their interconnection to the grid and the system operators, and what are the corresponding measures to it.

Keywords: Smart grid, renewables, distributed generations, renewable integrations, variability

Email: sarkari@nitrkl.ac.in

M. Kathiresh A. Mahaboob Subahani and G.R. Kanagachidambaresan (eds.) Integration of Renewable Energy Sources with Smart Grid, (101–118) © 2021 Scrivener Publishing LLC

5.1   Introduction

One of the effective way of utilizing the distributed energy resources such as wind, solar, geothermal, hydel power, and fuel cell is the smart grid technology [1–3]. With fast diminishing of fossil fuel energy resources and their emission effects on the environment, the search for alternate energy sources is the key concern the entire world is currently dealing with. However, with the advancement in technologies for integrating the renewable energy resources and the corresponding reduction in cost has created enormous opportunities to address the scope of electricity generation from renewable energy resources to great extent. In such scenarios, the smart grid technology plays an important role by providing superior quality power, coordinated generations, and options for energy storage to encounter the fluctuations and the operating challenges. It also plays a major role in promoting the consumer's involvement in decision making and providing the operating environment suitable for both the supplier and the consumer. For example, in smart grid system, the customers get the opportunity to supply power to the grid by using distributed generation sources like solar, wind, electric vehicle, or energy storages. Therefore, the reliability of the grid improves significantly using proper demand response plans, utilizing more distributed generation resources, and use of energy storing systems with appropriate power flow control. With this, the penetration of renewable energy resources (an illustration of penetration level is presented in Figure 5.1) to smart grid system is increasing to provide reliable and cheap electricity services.

5.1.1   Smart Grid

Conventional power system network has large power generating plants located at remote sites from the load centers (or consumers) and the generated power is transmitted over a large distance through overhead transmission lines to the load centers to distribute power to the customers

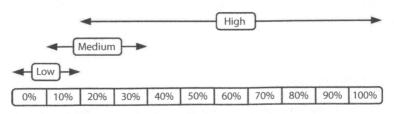

Figure 5.1 Penetration levels of distributed energy resources.

(Figure 5.2a). However, the scenario of today's world has changed with the introduction of renewable energy resources in the form of distributed generating units spreading over the power distribution network (Figure 5.2b). Apart from the environmental benefits, the use of distributed generations has the benefit of cheap electricity generation and transmission due to low transmission losses. However, the limitation of energy generation from renewable sources is its variable nature of power generation in daily and seasonal patterns. However, with the current state of the art in power electronics technology, the characteristics of distributed generation are matched with the grid integration requirements, and hence, the establishment of large distributed generation has now became feasible.

The two major technical challenges in high penetration of renewable energy generations are managing the variability and uncertainty in power generations and, most importantly, the supply-demand management during power contingencies [4]. Therefore, the smart grid technology is employed to mitigate the energy balancing and power flow issues, which includes:

- Skillful predictions of power generations, which allows system planner and operators to manage variability and uncertainty more effectively;
- Power electronic interfaces for automatic grid support;

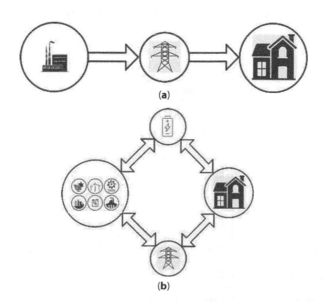

Figure 5.2 (a) Conventional power system network. (b) Smart grid network.

Figure 5.3 Basic structure of a smart grid system.

- Presence of smart-meter helps in control, metering, billing, and consumer contributions for efficient power balancing;
- Availability of energy storage systems (ESSs) to mitigate the short-term variations in renewable energy generations;
- Ability for real-time monitoring and managing.

Figure 5.3 represents the basic structure of a smart grid system, where unlike convention power system network, the power generations are from renewable energy resources like solar and wind apart from the conventional

Table 5.1 Comparing conventional grid with the smart-grid.

Parameters	Conventional grid	Smart grid
Electricity Generation	Conventional generations	Distributed generations and conventional generations
Pollution Level	High	Low
Efficiency	Low	High
Controllability	Low	High
System Monitoring	Low	Effective monitoring
System Network	Radial	Integrated
Data Communication	Unidirectional	Bi-directional

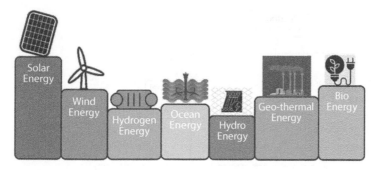

Figure 5.4 Renewable energy resources.

thermal power plants [1]. A comparison of conventional grid with smart grid is presented in Table 5.1.

5.1.2   Renewable Energy Resources

The list of key renewable energy resources is shown in Figure 5.4 and the distributed generation categories is presented in Figure 5.5 [5, 6]. The brief introduction to these energy resources are as follows:

1. Fuel cell: Fuel cell is an electrochemical device which generates electrical and thermal energy from chemical energy using electrochemical process similar to a battery; however, the difference is since the electrochemical resources used in fuel cell (hydrogen and oxygen) are available in abandon, there is no need for the continuous feeding of the raw materials.

A typical fuel cell process is shown in Figure 5.6, where the raw materials are H_2 and O_2, and the H_2O is the byproduct with generation of electrical energy in the form of dc power. With today's fuel cell technology, immobile fuel cells of mega-watt power ratings are available, whereas portable fuel cells of lower power ratings are also readily available. The fuel cell energy technology is termed as clean and green energy, as the electrochemical process involves hydrogen and oxygen, which are available in excess in nature, and their combination electrochemical process generates heat and electricity without any combustion process with low environmental concerns. Since the output electrical energy from a fuel cell is dc, hence single-phase or three-phase dc-ac power converters are needed to convert the dc power into ac to feed to the ac utility grid. The advantages of fuel cell technologies are as follows:

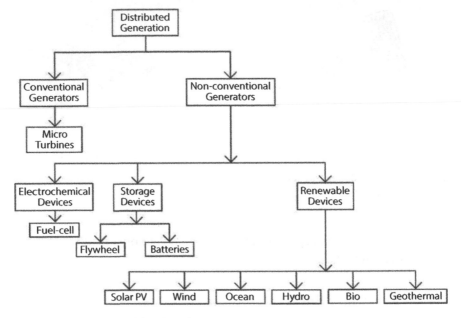

Figure 5.5 Categories of distributed energy resources.

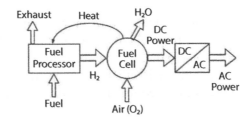

Figure 5.6 Typical fuel cell energy extraction process.

- Higher power conversation efficiency compared to conventional thermal power plants;
- Compact, less air pollutions, and low noise level;
- Modular design provides ease to match with the specific power demand.

One major disadvantage of fuel cell generator is the variable nature of internal impedance, i.e., the internal impedance of a fuel cell changes with time, and hence, there is a need for some external means to monitor and regulate the fuel cell terminal voltage.

2. Solar PV Energy: Solar energy is one of the most readily available renewable energy where the solar photovoltaic effect converts the solar energy into electrical energy. The radiations available from the sunlight are used to energize the solar panels which, in turn, generates unregulated dc power. This unregulated dc power is then converted into regulated dc with maximum power point tracking (MPPT) to extract maximum power and then converted into grid frequency ac using dc to ac power converter. A typical configuration of solar power extraction system is shown in Figure 5.7.

Extracting solar energy has lower environmental affects; available in abandon and modular design for flexible scale of installation are its key advantages, and capital investment, variability in power generations, and the low power conversion efficiency are its major disadvantages. However, steady researches are ongoing to mitigate the above-mentioned issues.

3. Wind Energy: In wind power extraction, the energy from the flow of wind, i.e., the kinetic energy of wind is extracted using large wind turbines, and electrical energy is generated using Permanent Magnet Synchronous Generators (PMSG) or Wound Rotor Induction Generators (WRIG) coupled with the turbine. Both the power extraction schemes are presented in Figures 5.8a and b, respectively. The electrical power delivered by a wind turbine-generator system depends on the wind velocity, the radius of turbine blades and the turbine speed characteristics. In today's wind power extraction technology, offshore wind farms are much popular for clean energy extraction than individual wind power installations [7, 8].

Apart from clean and green energy, the wind energy extraction has the disadvantages like uncertainty in availability of power and uncontrolled output power, i.e., very random and stochastic power output behavior.

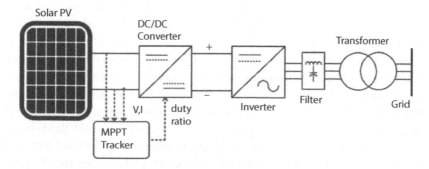

Figure 5.7 A typical configuration for solar power extraction.

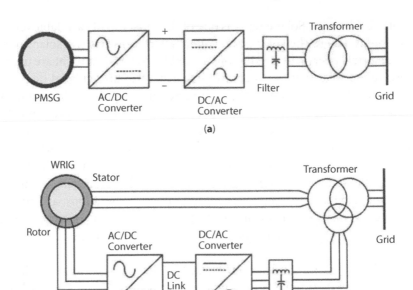

Figure 5.8 Two popular configurations for wind power extraction (a) using PMSG and (b) using WRIG.

Therefore, best practice is to use probability-based modeling techniques for optimum power allocations.

4. Hydropower: In hydropower extraction, the electrical power is generated from the potential and kinetic energy of running water (due to gravity) by building dams and water reservoirs. Hydropower extraction is an old power harnessing process, where running water is used to rotate the hydro turbines which then drive the salient pole synchronous generators or alternators to generate electrical power. Unlike coal-based power plants, hydropower plants have very high power conversion efficiency due to lower energy losses in power conversion process. Based on power generating capacity, the hydropower plants are categorized into large or small-scale power plants and the micro hydropower plants, and also based on power extraction approaches, like water dam based, running water based, and pumped storage based. In general, the conventional water dam–based large hydropower plants have much sluggish response to the load demand; however, the small and the micro hydropower plants have much quick response to the load changes and are easy to establish.

5. Ocean or Tidal Energy: Ocean or tidal energy is another category of renewable energy source due to their boundless availability and less environmental concerns. Several researches are going on (many in concept stages) to harness the ocean energy in an optimized and efficient technique and is realized as one of the key energy sources. In this energy conversion process, the kinetic energy of the ocean tides is converted into mechanical energy to drive the hydel turbines and electrical energy is generated from the generators coupled with the hydel turbine shaft. The power output of the turbine depends on tide velocity, density of water, thrust area, and the turbine efficiency. Therefore, more is the tide velocity and more is the thrust area, more will be the turbine power output. The tidal power is divided into the following categories.

- Outflow power generations where energy stored during peak is utilized during outflow;
- Power generations from flood;
- Two-way generations where power is generated by the turbines during both flood and tides.

The key limitation of ocean power extraction is the high initial cost and the massive sea power, which affects the installation and maintenance.

6. Bio Energy: Biomass is another widely accepted renewable energy source, where electrical power is generated from gaseous fuel from biomass or from the organic waste matters like trees and byproducts from agricultural processes. Various technologies are available to date to convert the biomass into gaseous fuel and specific technology is employed depending on the requirements. Apart from electric power generations, it is also used for transportation and cooking purposes. Generally, biomass plants are located to the locations where huge sources of biomass is available.

7. Geothermal Energy: Geothermal power extraction is comparatively newer energy harnessing process, where the heat energy present in the deep earth is utilized for energy conversion process. In this process, a fluid is circulated down into the deep earth to absorb the available heat and bring the same to the earth's surface. Therefore, fluids with higher specific heat capacity essentially used in this process. After accumulating the heat on the earth's surface, the same heat is used for heating and generating electrical energy similar to conventional thermal power plant. The only difference is in geothermal energy extraction process, the steam is generated using geothermal heat energy. Since, heating is done based on geothermal energy,

the cost of electricity production is much less and energy conversion efficiency is much has higher compared to conventional coal-based thermal power plants.

8. Energy Storage Systems: In smart grid technology, ESS plays the primary role in facilitating the integration of renewable energy resources or the distributed energy resources, which are variable and intermittent in nature [9]. It plays the role of spinning reserve as in conventional power system and hence eliminates the need for reserve power. Therefore, it enables efficient power transmission and distribution with reduced cost and less power failures. It helps in eliminating the fluctuations in power generations and enables integration of more renewables into the power system network. The ESS is classified as follows:

1. Mechanical: flywheel, compressed air, hydel;
2. Electrical: superconducting coil, super capacitor, battery storage system (BSS);
3. Chemical: fuel cell, BSS.

Out of these, super capacitor, battery, and flywheel are the widely used energy storing systems.

9. Electric Vehicles: Unlike conventional vehicle where internal combustion engines are employed for the vehicle propulsion, in electric vehicle, electric motor is used as the key component for propulsion. With the increase in cost of fossil fuel and scarcity in their availability, the concept of plug-in hybrid electric vehicles or fully electric vehicles are the alternative to conventional fossil fuel–based vehicles. In smart grid system, electric vehicles play a significant role while charging and while supporting the grid during grid contingencies, where power flows are either vehicle to grid and grid to vehicle. For charging of the electric vehicle, since more and more electric vehicles are introduced in the power system network, it became necessary to optimize the charging characteristics to minimize their effect on grid stability by not altering the peak power demand of the system. On the other hand, the power flow from the vehicle to grid is employed to mitigate the load-balancing problem due to supply fluctuations. Therefore, for quick response during charging and during contingency support, the suitable charging and discharging times are decisive for electric vehicle applications. Electric vehicle acts like a storage system in smart grid technology.

5.1.3   How Smart Grids Enable Renewables

The high penetration of renewable energy generations is only feasible by upgrading the conventional power system network to advanced solutions like smart grid technology to facilitate the integration of renewables of different characteristics [1]. For example, the renewable energy resources like solar, wind, ocean, and hydro are dependent on the climate conditions and hence their energy outputs are variable in nature. Since, in a power system network, the demand supply response need to be matched always (without storage systems), therefore either demand has to be met by the generations or need to store the excess energy generated in ESS to mitigate the power generation and/or demand fluctuations particularly for high penetrated distributed energy systems. Due to power generation variability from distributed energy generations, the utility companies are not comfortable to integrate such energy sources to the conventional utility grid due to concern over grid stability, safety, operations, and the pricing. Therefore, the smart grid technology is the seamless solution to address such challenges with added features, which further ease the transition from the conventional energy sources to renewable energy sources.

5.1.4   Smart Grid and Distributed Generation

The two promising distributed renewable energy resources are the solar and the wind, in particular the solar rooftop PVs, and with the use of smart grid technologies the same can be promoted to much extent. The use of smart grid technology trains the power system operators by providing continual real-time data about the system conditions and the control. Providing real-time data on distributed electrical generation enables the power system operators to manage the demand-supply response more effectively. For example, connecting or disconnecting the distributed energy generations as per the need or to match the load demand. The energy then generated from the disconnected energy generations is stored in the storage systems present in the smart grid system which further enables superior voltage regulations. Therefore, the smart grid technology provides the monitoring of the generations and load demand, which allows the utility companies to use the distributed renewable energy resources as an alternative to conventional fossil fuel–based power generations. It also features smart metering which enables the billing for the distributed energy generations [10].

5.1.5   Grid Integration Terminologies

Few frequently used grid integration terminologies are as listed below [11].

a. Area of Balancing or Balancing Area: It defines the area in which demand-supply balancing need to be maintained by the power system operators. The area includes power generation, transmission, distribution, and the consumer.

b. Capacity: Capacity is defined as the capacity of the power plant to cater the load demand and is defined either in watts or in percentage of nameplate rating.

c. Flexibility: Flexibility of a power system is defined as its ability to respond to the changes in power demand and in the supply.

d. Grid Integration of Renewable Energy Resources: It is defined as the efficient and cost-effective integration of the renewable energy sources to the grid while maintaining the grid stability and reliability.

e. Storage Systems: Storage systems in a power system network is defined as the capability of the system to store the electrical energy in case of surplus generations and utilizing the same at a later stage during any grid contingencies. Presence of storage system in a power system network significantly improves the system reliability.

f. Variable Renewable Energy: The electrical power generated from renewable energy sources like solar, wind, and hydro varies with climate changes and varies over time and hence not constant all over the day and all over the year. Therefore, the energy generated by them are variable in nature.

g. Variability: Fluctuations in load demand or/and power generations.

h. Uncertainty: Uncertainty in a power system is defined as the inability to predict the electricity supply and/or the demand.

5.2   Challenges in Integrating Renewables Into Smart Grid

Green earth is a global initiative today, which drives the quest for the alternate techniques for efficient power generations, transmission and distribution of electrical power. In such mission, one promising solution is the use of renewable energy resources for electrical power generations and integrating it with the existing power grid. However, considering financial and reliability constrains, such integration demand changes in the power system planning and its operations. As explained earlier, smart grid technology enables the efficient and effective integration of the renewables with the power grid and supports the reliable power system operations even with fluctuations in renewable power generations. In this integration process, the power converters like dc-dc, dc-ac and ac-dc converters plays the role of interfacing and helps effective power flow control.

The key factors to be considered during variable renewable energy integrations are the quality of output power, efficiency, ease of power conversion and integration, reliability, safety and most importantly the financial benefits. This section deals with the grid integration challenges and the requirements for flexible integration of the variable and uncertain renewables with the power grid [11].

5.2.1 The Power Flow Control of Distributed Energy Resources

In smart grid system integrated with distributed solar energy fed by the consumer, the consumer receives financial incentive for feeding power to the grid. Therefore, for financial benefits, the consumer always tries to maximize the PV power output and feed it to the grid; however, the same may not be acceptable by the utility supplier. The utility company may require to curtail the PV power based on the factors like stability, protection, cost, and many more. Therefore, there is a need for the power flow control to curtail the PV power output and the decision of which is a complex subject matter. Some of it can be resolved by contract agreement between the consumer and the utility supplier.

5.2.2 Investments on New Renewable Energy Generations

By considering the possibilities of grid integrations, the power system planners may extend their investments more on future generations based on renewable energy resources. Therefore, this, in turn, may further establishes vision to drive innovation in the strategies and measures to further support addition of new clean and sustainable energy generations. In addition to it, technologies that contribute more toward grid stability and system reliability like wind and solar energy systems will be on higher priority and need be motivated enough for further investments to overcome the negative influences of integrating variable resources to the grid. However, the difficulty in scaling up of variable renewable energy generations is the uncertainty in planning for long-term electricity demand, as with the presence of significant variable generations complicates the predictions and leads to improper system evaluations. Therefore, the power system planners and the operators need to follow some approach to evaluate and estimate the capacity of variable generations effectively, and optimal techniques need to be followed to utilize them economically.

The advantages of integrating distributed solar power over utility level solar power are reduction in transmission losses and no fuel cost for power generations. However, introducing distributed solar power introduces more challenges in system operations with chances in voltage variations due to increase in confined power generations and complicates the system operations due to unplanned power flows. Therefore, it is necessary to apprise the procedures, standards, and distribution-planning methodologies to align with the characteristics of the distributed energy generations for extracting the maximum benefits with or without grid support. System with increased distributed generations may experience slight fluctuations in power output which does not influence the power system operations. However, the slow variations in power generations results in increased need for spinning reserve and interventions to ensure stable power system operations. Such scenarios are much dependent on the share of renewable energy generations over the reserve capacity.

5.2.3   Transmission Expansion

Integrating more variable renewable energy resources to grid needs upgradation of the existing grid as well as its expansion to accommodate the change. Therefore, it enables the grid system to access more clean and green energy generations like solar, wind, and hydel from remote locations to the existing transmission networks. Hence, there is need for new strategies, guidelines, and measures to encourage new investments in transmission expansion and avoids the need for any changes in the power system networks.

5.2.4   Improved Flexibility

Flexibility is a key concern for the grid systems with high penetrated solar and wind power generations. Therefore, to improve the system flexibility, low cost becomes the key factor for implementing the scheduling, estimation, etc., to avoid the capital investments and to have considerable flexibility in the system. The same is further increased by practicing the coordinated operation of distributed energy generations and their scheduling. Furthermore, by using flexible transmission and flexible generations, the system flexibility can be further scaled up. Presence of ESS in smart grid system also helps in improving the flexibility of high-penetrated grid systems. However, accessing flexibility parameter is governed by the policy mechanisms for vertically integrated utilities, and incentives, market design mechanisms for restructured power markets.

5.2.5   High Penetration of Renewables in Future

For power system network, the planning is to access the long-term demand and evaluation for expanding its capacity and transmission. As mentioned earlier, with introduction of more distributed energy generations, the same need to be more focused to improve the power system flexibility. This, in turn, enables the power systems operators for reliable operation of grid system with high-penetrated distributed energy generations. The grid integration can be further studied based on past operation experiences to identify the potential constraints, which affects the system reliability and the potential measures to overcome the same. However, studies like economic dispatch, system-integrating cost of distributed energy resources like solar and wind, are comparatively much complex and challenging. Nevertheless, the same identifies the challenges and the complications that we may face while integrating wind and solar energy resources to the power system network and hence motivates for prioritizing and sequencing grid integration investments.

5.2.6   Standardizing Control of ESS

In smart grid system with integrated renewables, the ESS is the major component to ease out the power fluctuations as explained earlier. The interfacing of such storage systems are done through controlled power converters. The major contributions of ESS in smart grids are voltage and frequency stability, supply-demand matching, and improving system reliability and power quality. However, the key issue with such energy storing systems is the non-standardized control strategies since each of them needs different control technique unique for the specific ESS. Therefore, each then requires precise planning for specific control strategy, which, in turn, makes the complete system more complex and costly. Therefore, there is a need for standardizing the control strategies for the integrated ESS for ease of control and maintenance.

5.2.7   Regulations

Use of smart grid for integrating renewables has several advantages, which are difficult to outline, and hence, certain regulatory framework need to be devised to define the outcomes of smart grid technology. Moreover, with defined benefits of smart grid with integrated renewables, it ensures steady investments for its further expansion. It builds some confidence in smart grid operations as well. For example, when the power utility companies are much more confident about smart grid operations and its benefits with renewable integrations in terms of cost, they will come forward with more investments for smart grid solutions. This will only happen, when they are quite confident about the benefits of their investments in future smart grid with more penetration of distributed energy resources. Therefore, it explain the need for the regulations must be in place to upfront the investments and more returns to their investments and, in absence of such policies, will surely demotivate the future expansions.

5.2.8   Standards

System with no defined standards is not going to work satisfactorily. For example, standardizing and defining the voltage, frequency, etc., and defining the equipment sizes. The smart grid technology currently facing the same issue of lack of standards and several works are going on to defined a universal standard that defines the parameter levels and the way for several connections. The utilities may not be involved in the standardization process; however, it is recommended for them to align with the outcomes of the standard defining committee for emerging standards. The standard plays a vital role in integrating the new renewable energy resources unanimously.

5.3   Conclusion

With prospective significance of renewable energy sources in smart grid system, this chapter dealt with the key requirements and the challenges of integrating renewable energy resources to the smart grid system. First, some introduction to smart grid technology is stated in introduction and then some details about key renewable energy resources with their important benefits and limitations. In further section and subsections, the requirement aspects of grid integration of renewables are explained in details. It is concluded that, the best practice for integrating renewable energy resources to the smart grid system is to address most of the key challenges using best possible solutions. Such efficient integration surely will help in meeting electric power demands very effectively.

References

1. Kempener, R., Komor, P., Anderson, *Smart Grids and Renewables: A Guide for Effective Deployment*, International Renewable Energy Agency (IREA), UAE, 2013, https://www.irena.org/publications/2013/Nov/Smart-Grids-and-Renewables-A-Guide-for-Effective-Deployment.
2. Nawaz, M.K. and Zafar, S., Integration of Renewable Energy Sources In Smart Grid: A Review. *The Nucleus*, 50, 311–327, 2013.
3. Arul, I., *First International Conference on Large Scale Grid Integration of Renewable Energy in India*, New Delhi, September 2017.
4. Phuangpornpitak, N. and Tia, S., Opportunities and challenges of integrating renewable energy in smart grid system. *Energy Procedia*, 34, 282–290, 2013.
5. El-Khattam, W. and Salama, M.M.A., Distributed generation technologies, definitions and benefits. *Electr. Power Syst. Res.*, 71, 2, 119–128, 2004.
6. Khan, B. and Singh, P., Optimal Power Flow Techniques under Characterization of Conventional and Renewable Energy Sources: A Comprehensive Analysis. *J. Eng.*, 2017, 16 pages, 2017.
7. 1. Akbarali, M.S., Subramanium, S.K., Natarajan, K., Real and Reactive Power Control of SEIG Systems for Supplying Isolated DC Loads. *J. Inst. Eng. India Ser. B*, 99, 587–595, 2018, https://doi.org/10.1007/s40031-018-0350-8.
8. 2. Akbarali, M.S., Subramanium, S.K., Natarajan, K., Modeling, analysis, and control of wind-driven induction generators supplying DC loads under various operating conditions. *Wind Eng.*, 1–16, 2020, https://doi.org/10.1177/0309524X20925398.
9. Teleke, S., *Energy storage overview: applications, technologies and economical evaluation*, pp. 1–11, White Paper, Quanta Technology, 2010.
10. Dai, M., Nanda, M., Jung, J., Power flow control of a single distributed generation unit. *IEEE Trans. Power Electron.*, 23, 1, 343–352, 2008.
11. Cochran, J. and Katz, J., *Integrating Variable Renewable Energy Into The Grid: Key Issues*, National Renewable Energy Laboratory, USA, 2015, https://www.nrel.gov/docs/fy15osti/63033.pdf.

6

Grid Energy Storage Technologies

Chandra Sekhar Nalamati

Electrical Engineering Department, Motilal Nehru National Institute of Technology Allahabad, Prayagraj, India

Abstract

Ever increasing world energy consumption, limited global resources for tradition fossil fuel–based power supply networks, deregulated world energy markets, and global environmental concerns have been changed the global energy landscape from centralized power network to green renewable energy integrated distributed power network. Green clean renewable energy sources include solar photovoltaics, wind turbine based, combined heat and power (CHP), micro turbines, and fuel cells system. By 2040, the world will more rely on solar and wind supplies that are together expected to increase their share three times in the global energy. The complementary pattern regimes of the wind and solar PV sources coupled with the energy storage technology support makes them firm energy source. Energy storage technologies help in reducing power generation-consumption gap and also in improving grid power quality and stability. Most of the energy storage technologies are deep-rooted. Also, the advances in material and processing technologies and bulk production result in reduction of cost. The specific energy storage technology has been chosen based on need of power density, energy density, and cost for different durations (short-term or long-term) and applications.

The main objective of this chapter is to acquaint with overview on various grid electrical energy storage technologies with their detailed characteristics comparison and their applications.

Keywords: Grid energy storage, renewable power generation and smart grid

---

Email: chandu3072002@gmail.com

M. Kathiresh A. Mahaboob Subahani and G.R. Kanagachidambaresan (eds.) Integration of Renewable Energy Sources with Smart Grid, (119–140) © 2021 Scrivener Publishing LLC

6.1   Introduction

The traditional power supply system has been characterized by non-distributed power generation plants, one-way power flow direction from generation to the customers, fossil fuel–based generation, and passive nature. The world population and its growing energy demand generation, concerns regarding environmental pollution of fossil fuel–based power generations, and need of other power generation resources to fulfill the current energy demand have changed the conventional power system paradigm to a new structure. The possible ways to balance the generation and demand are installation of new power generation plants with corresponding transmission, distribution system, and using customer side demand response approaches and adding energy storage with green energy sources. However, first two solutions are not feasible due to economic issues and time-to-time dealing with customers. The viable solution for new power system structure allows the environmental friendly renewable energy resources integration along with the electrical energy storage system to help to fulfill the need of rising energy demand and growing environmental concerns. However, the environmental friendly renewable energy sources (RES) availability depends on their geographical location and also they are uncontrollable in nature. The electrical energy storage systems help RES to improve their efficient usage and make them better power sources [1–7]. Figure 6.1 shows the smart grid structure showing its role to help in

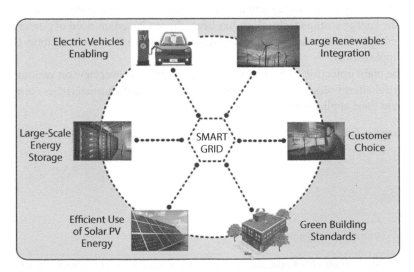

Figure 6.1 Smart grid with its enablers.

achieving efficient way of usage of energy sources along with advanced information and communication technology to solve the energy demand gap issue.

The enablers that make smart grid a promising solution are as follows [8–10]:

1. Advanced information and communication technology along with large on and off-field instrumentation that helps in good way of forecasting of variable RES.
2. Smart power electronics–based converter technology.
3. Advanced technology-based smart meters for demand side response.
4. Electrical energy storage integration to help RES and supply-demand gap.
5. Smart way of the system management with real time data.

One of the great global concerns is global warming and to maintain it well below the 2°C. To reach this target globally, there should be 266 Gigawatts of energy storage availability by the year 2030 and predicted that the world energy storage market reach approximately 942 Gigawatts of energy storage by the year 2040 [11–14].

6.1.1 Need of Energy Storage System

Above section highlights the role of energy storage technologies in light of issues concerning the power generation-demand gap, present and future energy security, climate change, and uncertainty in availability of continuous power supply to the end-energy customers. Figure 6.2 shows

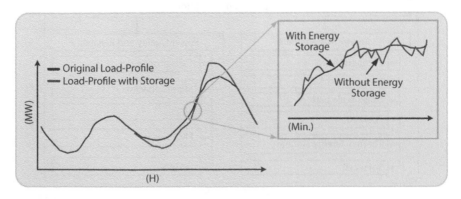

Figure 6.2 Importance of energy storage system.

the importance of energy storage system in enhancing whole system load profile. Another concern will be the efficiency of the power transmission of traditional power supply system from generating station to the end-energy customers. In the present era, the electrical energy storage system has gained tremendous momentum due to its capabilities including [5–7]:

1. Meeting energy demand requirements of short duration.
2. Efficient way of RES integration into the grid.
3. Ability to deal with the load leveling and peak shaving concepts.
4. Better energy security along with minimum adverse impact of environment.
5. Avoids new installations of fossil fuel–based power plants and corresponding reduction in transmission and distribution losses and cost.
6. Improved grid stability and reliability.
7. Whole system's enhanced operational performance.

6.1.2   Services Provided by Energy Storage System

Energy storage systems have been used for various applications depending on the requirements. In some applications, there is need of quick energy supply, whereas in some applications, power quality with energy supply is major concern and other applications require bulk storage. Figure 6.3 depicts the categorization based on the area of application of energy

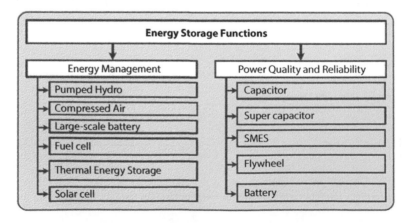

Figure 6.3 Functions of energy storage system.

storage technologies such as energy management and enhancement of power quality and reliability [15, 16].

6.2   Grid Energy Storage Technologies: Classification

Energy storage technologies can be categorized based on various methods or parameters as reported in the recent research work [17–20]. It can be based on their storage capacity, their functional characteristics, and their nature of response. The widely reported energy storage technologies categorization in the literature was based on the form of storage medium (Figure 6.4). Electrical energy may be efficiently stored in different available forms such as chemical and electrochemical energy, electric and magnetic field energy, potential and kinetic energy, compressed air, and thermal energy. A brief discussion on each storage technology has been presented in this section.

6.2.1   Pumped Hydro Storage System

This energy storage technology is the mechanical storage system and is one of the oldest global matured storage technologies that have large scale storage capacity. Figure 6.5 shows the detailed view of pumped hydro energy storage system having upper and lower (two) reservoirs, motor-generator set, and pump turbine along with power conditioning equipment. In low-demand

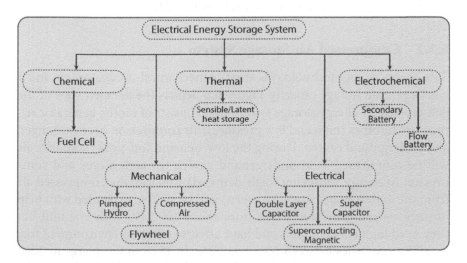

Figure 6.4 Energy storage technologies classifications.

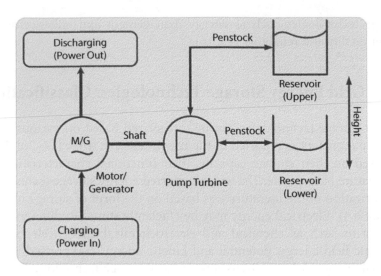

Figure 6.5 Pumped hydro energy storage system schematic.

scenario, energy will be stored through the process of sending water to the upper reservoir from the lower one with help of the motor and energy has been stored. In energy-retrieving scenario, the turbine and generator set helps to retrieve the electrical energy through the reverse process. Upper reservoir water volume and its height decide the amount of energy to be stored. The advanced developments helped to support variable speed operation. It also have high power and high energy profile with very long duration storage capacity. It needs high initial installation cost and suitable location also important to choose to obtain better results [20–22].

6.2.2   Compressed Air Storage System

This energy storage technology is also the mechanical storage system and it has large scale storage capacity. Figure 6.6 depicts the detailed view of the compressed air energy storage system that consists of motor-generator set, pressure turbines (high and low pressure), air compressor, heat exchanger, and underground cavern. During the low-demand (off-peak) scenario, air will be compressed in to underground cavern through motor and compressor. In energy-retrieving (high-demand) scenario, the compressed air from the underground storage (cavern) will be heated, expanded with help of turbines and generator. It can store above 100 MW with one unit but they need high energy inputs and have adverse effect on environment too. Its efficiency varies between 50% and 70%. In 1978 in Germany, maiden compressed air storage technology plant has been started [20, 23, 24].

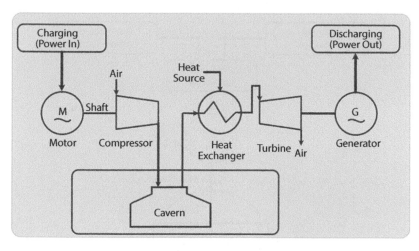

Figure 6.6 Compressed air storage diagram.

6.2.3 Flywheel Energy Storage System

The flywheel energy storage technology is also the mechanical storage system and it stores energy in kinetic form. Figure 6.7 depicts the detailed view of the flywheel energy storage system having flywheel, vacuum, and cooling pump along with power conditioning equipment, motor-generator set, and bearings. During the low-demand (off-peak) scenario, the system stores the energy in kinetic energy form with help of motor and rotation of flywheel. In energy-retrieving (high-demand) scenario, the electrical energy is retrieved from kinetic energy with help of flywheel and generator. Its efficiency varies between 80% and 85%. It has large power density as well as low energy density. The mass and rotational speed of flywheel will determine the storage capacity [20, 25, 26].

6.2.4 Superconducting Magnet Storage System

It comes under electrical storage system and stores the energy in magnetic field. Figure 6.8 depicts the detailed schematic diagram of superconducting magnet storage system that consists of superconducting magnet and coils, liquid helium/hydrogen, pump, vacuum insulated vessel, and refrigerator system along with power conditioning equipment.

The vacuum insulated vessel with liquid helium/hydrogen maintains the superconducting coil in the superconducting state. The coil self-inductance and double of current magnitude and coil operating temperature determines its storage capacity. The superconducting coil in the superconducting state has almost zero resistance and it helps in energy storage

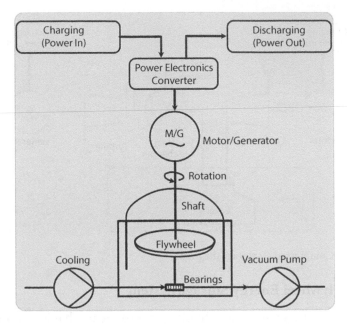

Figure 6.7 Detailed view of flywheel storage system.

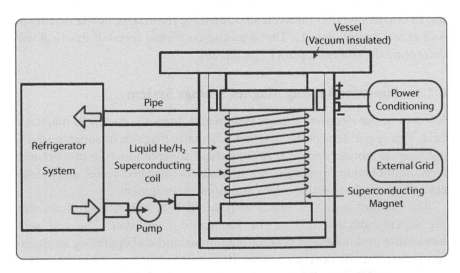

Figure 6.8 Superconducting magnet energy storage system schematic diagram.

with zero heat dissipation losses. Niobium-Titanium of having supercon-ducting critical temperature of 9.2 K is one of the good options. During the low-demand (off-peak) scenario, the system stores the energy in direct current form, whereas in energy-retrieving (high-demand) scenario, the electrical energy is retrieved with help of power conditioning system. Its efficiency is very high (around 98%) and responds very quickly. It is not economical and losses occur in refrigeration process [27, 28, 37].

6.2.5   Battery Storage System

Battery storage technology was one of the matured and oldest technolo-gies in the history. The advancements in the material and processing tech-nology over last two decades have made the batteries to be available in different ratings for different applications from watts (watt hour) to mega-watts (megawatt hours). They store energy in the form of electrochemical form. The basic structure of battery cell and its detailed internal structure have been depicted in Figure 6.9. It is of having two electrodes (one is positive potential anode and other is negative potential cathode) and an electrolyte (either in solid, molten state, or liquid state). They have good conversion efficiency of 65% to 90% that depends on battery cycle time and type of electrochemical reactions. Each battery cell has voltage capac-ity of few volts. In order to get high voltage and currents, battery cells are arranged in either series or in shunt way of connection. Battery stor-age technologies can be classified based on the electrochemical reaction that is reversible or non-reversible. The primary batteries are having non-reversible electrochemical reactions, whereas the secondary batteries are having reversible electrochemical reactions (rechargeable). Battery cell capacity has been usually indicated in ampere-hours (Ah). The balance energy in the battery is indicated as SOC (state of charge). It has been not

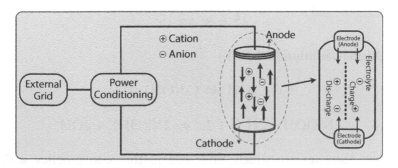

Figure 6.9 Battery storage system and detailed structure.

suggestible to overcharging and deep discharging. A good battery management helps in well control of charging and discharging process and extends the battery life.

There are different categories of batteries available in the present global market with different ratings for different applications [29–32]. The lead-acid batteries are the oldest technology and are most commonly used batteries but they have low energy density and also affect the environment. They are of secondary type and mostly preferred in vehicles and other applications that demand high range of load current. Whereas, lithium-ion (Li-ion) batteries are the most popular and have long life with very little self-discharge rate. They have large energy density and not result in any harm. To increase battery life, battery scheduled cycling or memory is not required. The chemical reactions at electrodes of various batteries have been given below.

Lead-acid (2.0V)

$$Pb + SO_4^{2-} \Leftrightarrow PbSO_4 + 2e^-$$

$$PbO_2 + SO_4^{2-} + 4H^+ + 2e^- \Leftrightarrow PbSO_4 + 2H_2O$$

Lithium-ion (3.7V)

$$C + nLi^+ + ne^- \Leftrightarrow Li_n C$$

$$LiXXO_2 \Leftrightarrow Li_{1-n}XXO_2 + nLi^+ + ne^-$$

Sodium-sulfur (~2.08V)

$$2Na \Leftrightarrow 2Na^+ + 2e^-$$

$$\lambda S + 2e^- \Leftrightarrow \lambda S^{2-}$$

Nickel-cadmium(1.0-1.3V)

$$Cd + 2OH^- \Leftrightarrow Cd(OH)_2 + 2e^-$$

$$2NiOOH + 2H_2O + 2e^- \Leftrightarrow 2Ni(OH)_2 + 2OH^-$$

Li-ion batteries are widely preferred in the auto mobiles such as electric vehicles due to their proved capabilities. These are also used in mobile phones, digital cameras, and laptops. Nickel-cadmium batteries also secondary type (rechargeable) and are temperature tolerant and have longer deep cycle but have memory effect. Applications are portable battery-operated devices, small computers and cam recorders. Sodium-sulfur battery is also known as liquid metal batteries which are made from sodium and sulfur. They show large energy density, good round trip efficiency (89%–92%), and made from inexpensive materials. This is also good option for the large energy storage applications. Flow batteries are treated as best substitute for the Li-ion batteries but they are not popular as Li-ion batteries. They have good power density with low energy density and have long life cycles. They can be preferred for continuous power supply applications. They also have good percentage is global battery market.

6.2.6   Capacitors and Super Capacitor Storage System

As discussed in the above section about the batteries, batteries store energy in the electrochemical form, whereas capacitors store energy in electrostatic field form that induced between two electrodes of the electrostatic capacitor. As there are no chemical reactions, capacitors induced charge can be easily reversible. Their storage capacity depends on the applied voltage, area of plates, and distance between the conducting plates. Compared to the batteries, the capacitor stores low electrical energy but quick to charge and discharge. They are preferred in applications such as power quality and power backup.

The super capacitors (SC) are also called as ultracapacitor (UC) or electric double-layer capacitors (ELDCs). Their detailed structure consists of a porous membrane separator, two conductor electrodes, and an electrolyte as depicted in Figure 6.10. They do not have direct dielectric material and stores energy in the double layer formed between surfaces of each electrode and the electrolyte. As results, ELDC result in better performance with higher value of capacitance and enhanced charge/discharge life cycle compared to the conventional capacitors. As a result, ELDC can give merits such as improved storage capacity than conventional capacitors and better charge/discharge cycle life compared to conventional capacitors. To enhance the capacitance of ELDC by increasing the electrode surface area, nanocarbon materials and graphene-based electrodes have been preferred. With the best designs, always ELDC efficiency has been more than 90% and very quick in response compare to both the battery technology and the conventional capacitors. However, their energy density is lower than the batteries [33, 34, 37].

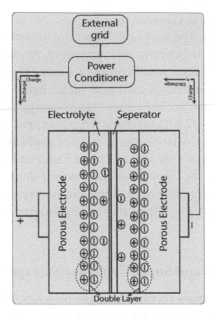

Figure 6.10 Super capacitor energy storage system.

6.2.7   Fuel Cell Energy Storage System

William Grove first introduced the operating principle of fuel cell principle in 1839. In 1960s, NASA has tested the first commercial expensive fuel cells for space satellite applications. From 1970s to till date, research and developments in the field of fuel cell technology has been very active and presented for small to large power applications. Various categories are solid oxide type, polymer electrode to membrane type, alkaline type, phosphoric acid type, direct methanol type, and molten carbonate type [35]. Hydrogen fuel cells are the popular among the all types. As depicted in Figure 6.11, hydrogen fuel cells consist of anode, cathode, catalyst, and membrane. Through electrochemical reactions (as shown below), it converts chemical energy into electrical energy, H_2O (water) and heat. Through the porous membrane (electrolyte), the hydrogen fuel and air reacts. As a results, there exist electrons and ions transfer across the electrolyte from the anode to the cathode.

The complete circuit has been formed when external load is connected and current flows through it. The water is the byproduct of this process and with the process of electrolysis, oxygen and hydrogen can be produced from the water. A single fuel cell can generate low power, so fuel cells can be connected in series to generate a standard voltage value for particular

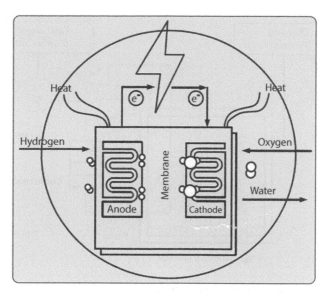

Figure 6.11 Fuel cell schematic diagram.

application. They are reliable and no moving machine parts and have high energy density. The usage of platinum makes them fuel cells expensive.

Anode:
$$2H_2 => 4H^+ + 4e^-$$
Cathode:
$$O_2 + 4H^+ + 4e^- => 2H_2O$$
Net Reaction:
$$2H_2 + O_2 => 2H_2O$$

6.2.8   Thermal Storage System

Thermal energy storage system depicted in Figure 6.12 consists of a high insulated storage tank with ohmic heater and heat exchanger, turbine, generator, condenser, pump, and power conditioning system. The high insulated storage tank environment will help to store the energy in the form of heat by heating water, salts, rocks, and other materials. In general, thermal storage systems are categorized into high temperature and low temperature storage systems. It is also classified as an industrial or building systems with either heating or cooling phenomenon depending on the different range of temperatures. In low-demand scenario, the high temperature heat can be obtained using ohmic heater from the electrical energy. In energy-retrieving scenario, heat will be obtained from thermal energy which

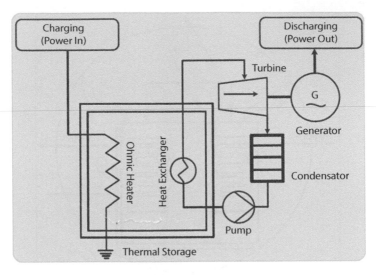

Figure 6.12 Thermal storage system schematic.

can be used to generate heat that can drive the turbine and the generator system to result in electrical energy. The peak heating requirements can be fulfilled with help of hot water storage system. They are matured and old technology, easy to store in bulk, and inexpensive with high reliability. Their round trip efficiency can be varied between 50% and 90% based on thermal storage type preferred [20, 36].

6.3   Grid Energy Storage Technologies: Analogy

The technical maturity of different energy storage technologies has been depicted in Figure 6.13. Some of them are in the matured stage, some in deployment and demonstration and development stage, and few are in active research stage. The pumper hydro storage, compresses air, and lead acid battery technologies are matured storage technologies among the all. UCs, super conducting magnet storage, flywheel, and few battery storage technologies are in emerging way. The compressed air and pumped hydro mechanical storage technologies are the preferred storage technologies for the applications with the large scale energy storage requirement [19, 37–40].

Energy density (expressed in Wh/L) and power density (expressed in W/L) are two significant parameters that gives the information about the storage capacity and charge/discharge response at which power can be released per unit volume. These two significant parameters play a vital role in choosing energy storage technologies for different applications. The

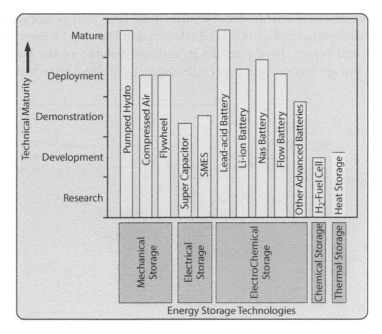

Figure 6.13 Technical maturity of energy storage technologies.

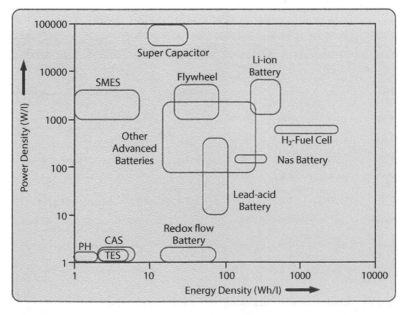

Figure 6.14 Power and energy densities of different energy storage technologies.

power density and energy density varies with each energy storage technology as depicted in Figure 6.14. As depicted in Figure 6.14, large energy density and power density results in reduced volume of the particular energy storage technology. Super/ultracapacitor technologies have more

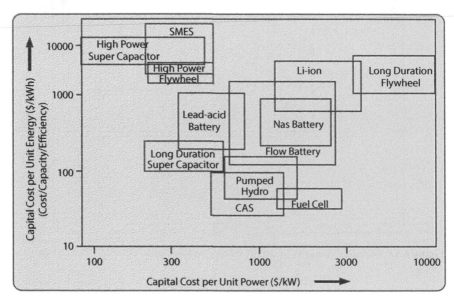

Figure 6.15 Capital cost–based comparison.

Types of Energy Storage	Specific Power (W/kg)	Specific Energy (Wh/kg)	Cycle life (Cycles)	Self-discharge at 25°C per month	Efficiency (%)	Typical Applications
Super Capacitor	2000-14,000	1.5-15	10^4-10^6	Very low	>90	Power Quality, Effective Connection, Voltage Control
Lead-acid Battery	100-200	20-40	200-2500	Medium	70-80	Off-Grid, Emergency supply, Time shifting, Power quality, Residential storage system
Li-ion Battery	300-1500	100-300	2000-5000	Low	80-90	Power Quality, Network efficiency, Off-Grid, Time shifting, Electric vehicle, UPS
NiMH Battery	220-1000	60-120	500-2000	High	50-80	Electric vehicle
NaS Battery	150-230	150-240	2000-4500	Very low	75-90	Time shifting, Network efficiency, Off-Grid
ZnBr Flow Battery	300-600	30-60	2000-3000	Very low	70-80	Long term storage
Flywheel	1000-5000	10-50	10^5-10^7	Very high	80-90	Power quality, Load levelling
Pumped Hydro	N/A	0.3-30	>20 years	Very low	65-80	Time shifting, Power quality, Emergency supply, Frequency control
Compressed Air	N/A	10-50	>20 years	High	50-70	Time shifting, Voltage control
Superconducting Magnet	500-2000	0.5-5	10^5-10^6	Very low	75-80	Time shifting, Power Quality, UPS

Figure 6.16 Detailed comparisons along with specific applications.

power densities and have small energy densities. The compressed air type, thermal energy storage, and pumped hydro type have smallest power and energy densities. The more energy densities with moderate power densities can be observed with fuel cell storage. Li-ion battery technology have better energy density as well as power density that makes them suitable for various range (small to large) of applications [19, 37–40].

The success of any energy storage technology directly influenced by the capital cost or investment for any selected application. The cost per unit energy is expressed as $/kWh, whereas $/kW is expression of cost per unit power. Figure 6.15 shows the detailed information about capital cost per unit power and energy of various storage technologies super/ultracapacitor technologies have low capital cost per unit power and more capital cost per unit energy. The compressed air, pumped hydro, and fuel cell storage technologies are with smallest capital cost per unit energy. Whereas, fuel cell technology has more capital cost per unit power. Super conducting magnetic storage can be seen with more capital cost per unit energy. Figure 6.16 depicts detailed comparison of energy storage technologies along with their specific applications [19, 37–40].

6.4 Applications of Energy Storage System

Even though electrical energy storage technologies are not primary energy sources, they can help in betterment of the power system performance characteristics. To avail smooth power flow operation with the uncontrollable

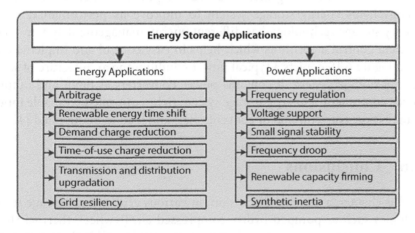

Figure 6.17 Applications of energy storage systems.

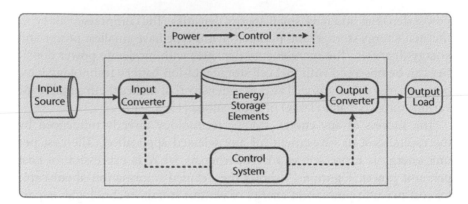

Figure 6.18 Power conditioning of energy storage system.

renewable energy sources, energy storage technologies are needed to incorporate within the system. Energy storage technologies helps at all levels of power supply chain such as at generation, transmission, and distribution. Figure 6.17 depict various applications of energy storage technologies that help in improving power quality, reliability, and stability, enhancing transmission system capability, and to have better power supply. Basically, applications are classified based on storage capacity (energy applications), response time, and real time control (power applications) [41].

6.5   Power Conditioning of Energy Storage System

The power conditioning system based on power semiconductor technology helps energy storage system to improve its performance. Each energy storage technology has different energy management system with efficient control algorithms which helps in easy control and implementation. Figure 6.18 depicts a typical power conditioning architecture of such energy storage system. It contains power converters at input and output ports, controller for energy storage system to be controlled, available input source (non-conventional or conventional) and load to be supplied [42].

6.6   Conclusions

In this chapter, a detailed discussion on various grid energy storage technologies such as pumped hydro, compressed air, flywheel, thermal storage, superconducting magnets, battery storage, SCs, and hydrogen storage have been presented. Necessity of energy storage technologies and their functions have been addressed. A comprehensive comparison of all energy storage technologies in terms of technical maturity, power and energy densities, and capital cost investment required per unit power and per unit energy have summarized. Applications and power conditioning of energy storage system are also presented.

The fast transformation of power system toward the futuristic power system structure with large clean renewables integration, electric vehicles, and local rural electrification concept has made storage as unavoidable power network component. Renewable adoption into grid without considering effectiveness of storage will results in great challenges. Day-to-day advancements in the storage technologies also makes difficulty in confirming particular storage technology as best for every application. Even though compressed air and pumped hydro technologies are matured and preferred for large energy storage, they need more space and depend on local conditions. Whereas, batteries, SCs, and fly wheel are available at any time. SC technologies have more power densities with favorable efficiency. With comparison with matured lead-acid batteries, Li-ion battery technology has better energy density as well as power density that makes them suitable for various range (small to large) of applications.

References

1. *Renewable capacity statistics 2020*, International Renewable Energy Agency (IRENA), Abu Dhabi, UAE, 2020.
2. *Sector Coupling in Europe: Powering Decarbonisation, Potential and Policy Implications of Electrifying the Economy*, Bloomberg New Energy Finance (BNEF), New York, 2020.
3. Bailera, M., Lisbona, P., Peña, B., Romeo, L.M., The Role of Energy Storage and Carbon Capture in Electricity Markets, in: *Energy Storage*, Springer, Cham, 2020.
4. Koohi-Kamali, S., Tyagi, V.V., Rahim, N.A., Panwar, N.L., Mokhlis, H., Emergence of energy storage technologies as the solution for reliable operation of smart power systems: A review. *Renew. Sustain. Energy Rev.*, 25, 135–165, Sep. 2013.
5. Del Rosso, A.D. and Eckroad, S.W., Energy storage for relief of transmission congestion. *IEEE Trans. Smart Grid*, 5, 2, 1138–1146, March 2014.
6. Chidanand Robert, F., Singh Sisodia, G., Gopalan, S., A critical review on the utilization of storage and demand response for the implementation of renewable energy microgrids. *Sustainable Cities Soc.*, 40, 735–745, 2018.
7. Rugolo, J. and Aziz, M.J., Electricity storage for intermittent renewable sources. *Energy Environ. Sci.*, 5, 5, 7151–7160, 2012, (IEEE).

8. Nourai, A. and Kearns, D., Smart grid goals realized with intelligent energy storage. *IEEE Power Energy Mag.*, 8, 2, 49–54, 2010.

9. Zame, K.K., Brehm, C.A., Nitica, A.T., Richard, C.L., Schweitzer III, G.D., Smart grid and energy storage: Policy recommendations. *Renewable Sustainable Energy Rev.*, 82, 1, 1646–1654, Jul. 2018.

10. Roberts, B.P. and Sandberg, C., The role of energy storage in development of smart grids. *Proc. IEEE*, 99, 6, 1139–1144, June 2011.

11. *Global Renewables Outlook: Energy transformation 2050*, International Renewable Energy Agency (IRENA), Abu Dhabi, UAE, 2020.

12. *Rivalry: the IHS Markit view of the energy future (2018-2050)*, IHS Markit, London, Jul. 2018.

13. *World Energy Outlook 2019*, Analysis, International Energy Agency (IEA), Paris, 2019.

14. *Share of wind and solar in electricity production, Global Energy Statistical Yearbook 2019*, Enerdata, France, 2019.

15. Palizban, O. and Kauhaniemi, K., Energy storage systems in modern grids-Matrix of technologies and applications. *J. Energy Storage*, 6, 248–259, 2016.

16. Yao, L., Yang, B., Cui, H., Zhuang, J., Ye, J., Xue, J., Challenges and progresses of energy storage technology and its application in power systems. *J. Mod. Power Syst. Clean Energy*, 4, 4, 519–528, Oct. 2016.

17. Rufer, A., *Energy Storage: Systems and Components*, 1st Ed., CRC Press, Boca Raton, Florida, USA, 2017.

18. Molina, M.G., Grid energy storage systems, in: *Power Electronics in Renewable Energy Systems and Smart Grid*, B.K. Bose (Ed.), Wiley, New York, NY, USA, 2020.

19. Nadeem, F., Hussain, S.M.S., Tiwari, P.K., Goswami, A.K., Ustun, T.S., Comparative Review of Energy Storage Systems, Their Roles, and Impacts on Future Power Systems. *IEEE Access*, 7, 4555–4585, 2019.

20. Dekka, A., Ghaffari, R., Venkatesh, B., Wu, B., A survey on energy storage technologies in power systems. *Proc. 2015 IEEE Electrical Power and Energy Conference (EPEC)*, pp. 105–111, 2015, London, UK.

21. Breeze, P., Pumped storage hydropower, in: *Power system energy storage technologies*, pp. 13–22, Academic Press, Elsevier Science Ltd, Amsterdam, 2018.

22. Rehman, S., Al-Hadhrami, L.M., Alam, M.M., Pumped hydro energy storage system: A technological review. *Renewable Sustainable Energy Rev.*, 44, 586–598, 2015.

23. Wang, J., Lu, K., Ma, L., Wang, J., Dooner, M., Miao, S., Li, J., Wang, D., Overview of Compressed Air Energy Storage and Technology Development. *Energies*, 10, 991, 2017.

24. Breeze, P., Compressed air energy storage, in: *Power System Energy Storage Technologies*, pp. 23–31, Academic Press, Elsevier Science Ltd, Amsterdam, 2018.

25. Alami, A.H., *Flywheel storage systems*, pp. 35–49, Springer International Publishing, Cham, 2020.

26. Arani, A.K., Karami, H., Gharehpetian, G., Hejazi, M., Review of flywheel energy storage systems structures and applications in power systems and microgrids. *Renewable Sustainable Energy Rev.*, 69, 9–18, 2017.

27. Ali, M.H., Wu, B., Dougal, R.A., An overview of SMES applications in power and energy systems. *IEEE Trans. Sustain. Energy*, 1, 1, 38–47, Apr. 2010.

28. Buckles, W. and Hassenzahl, W.V., Superconducting magnetic energy storage. *IEEE Power Eng. Rev.*, 20, 16–20, May 2000.

29. Linden, D. and Reddy, T.B., *Hand book of Batteries*, 3rd ed., McGraw-Hill, New York, NY, USA, 2002.

30. Moseley, P.T. and Garche, J., *Electrochemical Energy Storage for Renewable Sources and Grid Balancing*, Newnes, Elsevier Science Ltd, Amsterdam, 2015.

31. Hu, X., Zou, C., Zhang, C., Li, Y., Technological developments in batteries: A survey of principal roles, types, and management needs. *IEEE Power Energy Mag.*, 15, 5, 20–31, Sep./Oct. 2017.

32. Menictas, C., Skyllas Kazacos, M., Lim, T.M. (Eds.), *Advances in Batteries for Medium and Large scale Energy Storage: Types and Applications*, Woodhead Publ., Cambridge, 2015.

33. Conway, B.E., *Electrochemical Supercapacitors: Scientific Fundamentals and Technological Applications*, Kluwer Academic/Plenum Publishers, New York, 1999.

34. Sahay, K. and Dwivedi, B., Supercapacitors energy storage system for power quality improvement: An overview. *Quality*, 5, 4, 1–8, 2009.

35. Mekhilef, S., Saidur, R., Safari, A., Comparative study of different fuel cell technologies. *Renew. Sust. Energy Rev.*, 16, 1, 981–989, 2012.

36. Enescu, D., Chicco, G., Porumb, R., Seritan, G., Thermal Energy Storage for Grid Applications: Current Status and Emerging Trends. *Energies*, 13, 340, 2020.

37. Luo, X., Wang, J., Dooner, M., Clarke, J., Overview of current development in electrical energy storage technologies and the application potential in power system operation. *Appl. Energy*, 137, 511–536, 2015.

38. Sabihuddin, S., Kiprakis, A.E., Mueller, M., A numerical and graphical review of energy storage technologies. *Energies*, 8, 172–216, 2015.

39. Hamidi, S.A., Lonel, D.M., Nasiri, A., Modeling and Management of Batteries and Ultra capacitors for Renewable Energy Support in Electric Power Systems–An Overview. *Electr. Power Compon. Syst.*, 43, 12, 1434–1452, 2015.

40. Hadjipaschalis, I., Poullikkas, A., Efthimiou, V., Overview of current and future energy storage technologies for electric power applications. *Renewable Sustainable Energy Rev.*, 13, 1513–1522, 2009.

41. Opening the Box of Energy Storage. *IEEE Electrif. Mag.*, 6, 3, 8, Sep. 2018.

42. Molina, M.G., Energy Storage and Power Electronics Technologies: A Strong Combination to Empower the Transformation to the Smart Grid. *Proc. IEEE*, 105, 11, 2191–2219, Nov. 2017.

7

Multi-Mode Power Converter Topology for Renewable Energy Integration With Smart Grid

M. Sathiyanathan[1]*, S. Jaganathan[2] and R. L. Josephine[3]

[1]Department of EEE PSG Institute of Technology and Applied Research Neelambur, Coimbatore, India
[2]Department of EEE Dr. N.G.P. Institute of Technology and Applied Research Kalapatti, Coimbatore, India
[3]Department of EEE National Institute of Technology, Tiruchirappalli, India

Abstract

At present, the integration of various distributed energy resources is a key concept in the formation of smart grid. Particularly, solar and wind generators plays a major role in renewable power generation. But, these systems have many challenges to produce power at higher efficiency. The main challenge in solar photovoltaic (SPV) system is low cell efficiency. In addition, solar panel output is affected due the climate conditions such as Sun's irradiation level and temperature. Similarly, the output power from wind energy conversion (WEC) system is dependent on wind velocity. Instead of dealing with the climate change and low efficiency of a power source, the available power generated can be controlled and regulated using power converters to supply grid. Integration of solar system, wind energy, and other renewable energy sources to the smart grid provides a solution for above-mentioned challenges. The individual and co-generation of power from all the RES can be monitored and controlled through a Multi-Mode Power Converter (MMPC) which performs battery charging and inverter mode operation using single converter topology. This chapter presents detailed study of the MMPC in grid connected renewable energy application and comparative results with conventional converter control features are presented.

Keywords: Solar PV, multi-mode converters, smart grid, controller

*Corresponding author: sathiyangm@gmail.com

M. Kathiresh A. Mahaboob Subahani and G.R. Kanagachidambaresan (eds.) Integration of Renewable Energy Sources with Smart Grid, (141–170) © 2021 Scrivener Publishing LLC

7.1   Introduction

At present, the demand for renewable energy sources (RES) like solar photovoltaic (SPV) generators and wind turbine generators (WTGs) is being increased to meet the industrial and domestic power requirement and provides solution for an environment issues like greenhouse effect. The main drawback of RES is that the power generation depends on environmental conditions: SPV depends in solar lighting and temperature, and WTGs depends on wind velocity. Thus, the unregulated power from RES should be regulated before use. The connection between the output power from RES and consumer loads is classified into four categories. (a) Power direct: The solar PV/wind generator is directly connected to the direct current (DC) loads. (b) Stand-alone or Off-Grid: It is independent from the power grid where the batteries are used to store energy The battery power is utilized during absence of power from RES. (c) Grid Integrated System: This is also known as grid-tied system which consists of power source, power converters for regulations, and the load. This system has an option of using battery as a secondary source. (d) Hybrid: This system combines multiple RES to maximize availability of power. The hybrid grid connected system is most reliable solution to ensure the power generation and usage capacity. But, the development of modern power converters with modern technologies in Internet of Things (IoT), digital controllers, and highly intelligent control algorithms will ensure the grid connected operation. Moreover, the grid integration is the process of developing an efficient ways to deliver renewable energy power to the grid. Robust integration methods and IoT technology will maximize the cost-effectiveness of incorporating RES into the power system while maintaining or increasing system stability and reliability. The smart grid technology is a new era in electric power system which provides an unprecedented opportunity to the energy industry to provide reliable and more efficient power supply to the consumers. This smart grid technology will definitely support the power suppliers and consumers, particularly industries like automotive, manufacturing plants to speed up their process.

The international renewable energy agency (IRENA) stated that there are 2 million electric vehicle (EV) users in the year 2016, which will increase up to 100%, i.e., 200 million in 2030 worldwide mentioned in [1]. There are two types of EV in the market: battery operated vehicle and plug-in hybrid electric vehicle (PHEV), which uses both liquid fuel and battery. IRENA also stated that operation of EV or PHEV is highly relied on batteries and charging stations. Again, aforesaid issues such as unregulated power from RES, energy storage in battery, and requirement of charging stations are addressed only if the power sources and related system components are intelligent to handle the power requirement. These problems can be addressed by the recent developments in low and high power large-scale integrated circuits (ICs). The high power semi-conductor ICs used to build compact and cost effective power converters for the control and conversion of power from RES.

Particularly, in battery-powered operation, use of the battery at the lowest possible supply voltage is preferable for the longer operation. The battery's terminal voltage varies significantly depending on the state of its charging condition. Even electronic circuits in portable devices require both step-down and step-up functions of a power converter. The use of multiple power converters in the system will increase the size of the system and integrating or controlling all the converters by a single-controller is difficult in terms of coordination and also in the view of power quality standards. Therefore, a single and portable power converter is required to perform multiple operations which use H-bridge to perform AC-DC, DC-DC conversion for battery charging or driving direct DC loads, and DC-AC inverter operation. A novel multi-mode power converter (MMPC) and its control system are designed considering the above-mentioned aspects of grid connected operation. The MMPC system uses less number of discrete components such as semiconductor switches, inductor and capacitors in the circuit. The generation of suitable pulse width modulation (PWM) for DC-DC conversion is performed by modified perturbs and observes (MPO) maximum power point tracking (MPPT) algorithm and execution of inverter operation is performed using SPWM generation. This chapter proposes a new control method where Digital PWM is used for controlling converter. The RES and grid is monitored and suitable PWM control signal is generated using a FPGA controller. The FPGA is a low-cost device which provides parallel processing of multiple resources and features and it can be reprogrammed many times during the operation. This gives the user or operator to change the control feature and maintain the stability and transient performance of the converter over the wide range of the input voltage with keeping higher conversion efficiency.

The chapter is organized as follows. Section 7.2 presents literature survey where the problems associated with RESs, and various power converters design challenges are studied and presented. Section 7.3 describes the system architecture and their characteristics. Section 7.4 presents the MMPC modes of operation. Section 7.5 elaborates the implementation of control scheme of MMPC using FPGA. Finally, Section 7.6 discusses the simulation and hardware results.

7.2   Literature Survey

IRENA stated that the use of RESs may double in the year 2030 worldwide which is highlighted in [1]. Similarly, the requirement of efficient power converters increases simultaneously to automate the power generation and distribution, and many researchers presented the power converter topologies and their control strategies in the field of RES and grid integration. The buck-boost converter for WTG application in presented in [2, 3]. A buck converter is a DC-DC converter that serves to lower the voltage. The changing wind speed causes the generator output voltage to fluctuate. The problem can be solved by buck converter and its control. The buck converters with high conversion efficiency are widely used in ultralow-power applications. Similarly, a control scheme for achieving improved efficiency in an H-bridge buck-boost DC-DC converter is discussed in [4–6]. The H-bridge buck-boost converter with a conventional control scheme operates at lower efficiency compared to buck and boost converter for the same step-down/up action, due to the absence of direct power component. A new control method called single voltage feedback and dual feed-forward control for H-bridge buck-boost converter is presented in [7].

In [8, 9], a novel adaptive delta modulation (ADM) control scheme is proposed as a driver for an interleaved DC-DC power converter, analysis of the converter is carried out to achieve the required system stability and to improve the dynamic performance. The proposed adaptive control scheme is verified for interleaved DC-DC synchronous buck converter system. The converter has two pairs of the switch, one acts as a synchronous buck and another as a synchronous boost operation, and therefore is suited for battery-powered applications to provide an output voltage that is below or above the input voltage.

Similarly, the papers [10, 11] introduce a new single-phase boost inverter for photovoltaic applications. The introduced inverter merges a boost converter and a single-phase inverter in a single-phase single-stage inverter in [12, 13]. The proposed inverter has a simpler topology and uses the minimum number of power electronic components. A proportional-integral (PI) controller is applied to the system to control the input AC in [14–16]. Moreover, carrier phase-shifted sine pulse width modulation (CPS-SPWM) technology is easy to implement, and the equivalent carrier frequency is high, which has become the most widely used modulation method of cascaded multilevel inverter. Similarly, the highly efficient variable step-size MPPT algorithm is presented in [17, 18] for stand-alone SPV battery charging system.

A single controller design based on DSP and FPGA is presented in [19, 20] which are suitable for paralleling multiple inverters. A simple synchronization scheme between DSP and FPGA based on universal parallel port (UPP) is proposed to eliminate the synchronization delay among inverters, and independent control of each converter can also be implemented. Similarly, a novel MMPC has been proposed with a simple switching structure which is formed by the reduction of device count as compared to individual conventional converters. Further, it will give the most economic operation with less complexity and compact in size. This converter can be configured as dc-dc, dc-ac, and ac-dc conversion as presented in [21, 22].

Considering the above-mentioned aspects, a novel MMPC and its control system is designed and developed using dsPIC30F4011 for low-power application in [23, 24]. The comparative study has been presented with various PWM methods and control algorithms. Even though, the proposed control system is efficient for MMPC control which has more number of discrete component in the circuit and it took a long time for MPPT algorithm execution/computation and SPWM generation. There is no option for the coder to reconfigure the control functions fully or partially during the execution. The attempt has been made to address above-mentioned issues using recent developments in VLSI technology called Field Programmable Gate Arrays (FPGAs).

7.3 System Architecture

The grid connected MMPC consists of solar PV array and wind energy generator as a primary source of energy and battery as secondary source. The basic structure of the proposed system is illustrated in Figure 7.1. The converter system is configured in such a way to operate four-modes. They are buck, boost inverter and rectifier.

The voltage and current sensors are used to monitor the power sources and also the grid line for the synchronized operation. The controller selects the power sources based on the availability of voltage and current at the terminals, for an example, if the solar PV voltage is higher than the wind generator, then the solar PV is connected to the MMPC through bi-pass switches. The power from solar PV or wind generator is regulated in MMPC to charge the battery and also feeds power to the grid. The various modes of MMPC are presented in detail in the subsequent sections.

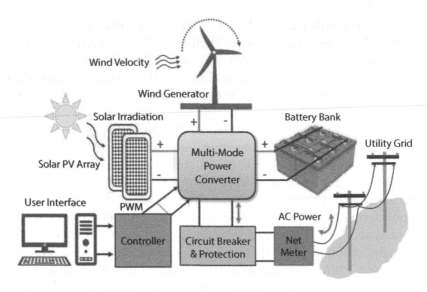

Figure 7.1 Block diagram of the proposed system.

7.3.1 Solar PV Array

The output characteristic of a photovoltaic panel depends on change in environmental conditions such as illumination intensity and temperature. The output characteristic has a unique maximum power point (MPP) at which the current-voltage product is greatest. The monthly radiation data in kWh/m² at Coimbatore, Tamilnadu, India is shown in Figure 7.2. The latitude and longitude detail of the aforesaid place are exactly 11.0254°N and 77.1246°E, respectively.

It is found that Coimbatore receives an average insolation energy level of 5 to 5.4 kWh/m²/day which is also verified from Global Horizontal Irradiation Map as per [25, 26]. This place receives total sun hour of 5 kWh/m²/day and 1,825 kWh/m² per year. This is highly sufficient to generate the required power from the solar PV system.

As mentioned earlier, the power generation of a solar panel per day may subject to change. Figure 7.3 illustrates the changes in output current (I) vs. voltage (V) and power (P) vs. current (I) at standard testing conditions (STCs), i.e., irradiation level of 1,000 W/m² and temperature of 25°C. Thus, the PV system should have a MPPT controller to control or track the maximum power. The purpose of MPPT algorithm is to ensure the power flow from to the load/energy storage device at highest level of efficiency. The MPPT controller typically controls the step-up or step-down DC-DC converter. This will regulate the voltage and current at the load terminal.

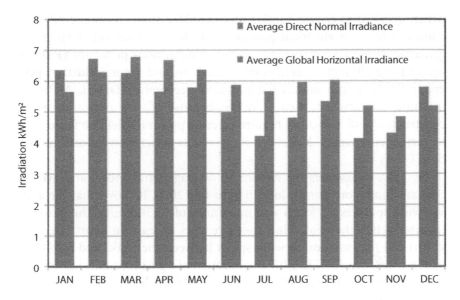

Figure 7.2 Monthly irradiation in Coimbatore.

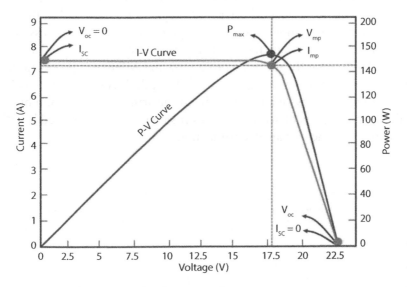

Figure 7.3 I-V and P-V curve of a solar PV panel.

7.3.2   Wind Energy Generator

The electrical generator is a rotational machine that converts the mechanical energy produced by the rotor blades into electrical energy. This energy conversion is based on Faraday's laws of electromagnetic induction, which is called as electro-motive force (EMF). There are many different configurations for electrical generators such as permanent magnet DC generator (PMDC), permanent magnet synchronous generator (PMSG), squirrel-cage induction generator (SCIG), wound rotor induction generator (WRIG), and doubly-fed induction generator (DFIG). A PMDC machine is extensively used in WTGs because of its simple and rugged construction and it does not require external power for field excitation due to permanent magnet. Moreover, it can use as either conventional motors or as DC WTGs.

Figure 7.4 illustrates the maximum, average wind speed, and average gust at Coimbatore during July 2016 to July 2020. It shows that the average wind speed is 6.2 km/h and maximum of 13.36 km/h. The gusts are reported when the peak wind speed reaches at least 16 knots (i.e., 29.6 km/h) and the variation in wind speed between the peaks and respites at least 9 knots (i.e., 16.65 km/h). The duration of a gust is usually less than 20 seconds. This wind speed will sufficient to generate higher electrical power from PMDC.

A PMDC machine consists of two parts: a stator and an armature. Here, the stator which is a steel cylinder. The magnets are mounted in the inner periphery of this cylinder and the field poles of this motor are essentially made of permanent magnet. If rotor, i.e., magnetic field, is rotated at higher wind velocity, then the machine producing an EMF that is proportional to its speed of rotation and magnetic field strength. Electrical power as indicated in Figure 7.5 is taken directly from the armature via carbon brushes with the magnetic field, which controls the power, being supplied by either permanent magnets or an electromagnet.

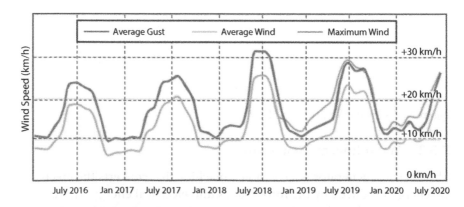

Figure 7.4 Wind speed at Coimbatore.

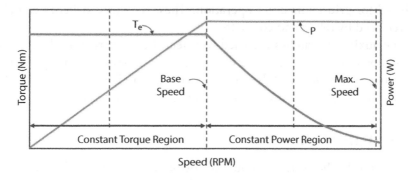

Figure 7.5 Speed vs. torque and speed vs. power curve of PMDC generator.

7.4 Modes of Operation of Multi-Mode Power Converter

The MMPC consists of nine semiconductor switches with SPV panel, V_{s1}, PMDC generator, V_{s2}, as an input sources and battery, V_{Bat}, as an energy storage device. Among nine power switches as illustrated in Figure 7.6, five switches, S_1–S_5 are considered as main switches and remaining Q_1–Q_4 are considered as auxiliary switches. The switch Q_1 and Q_2 are associated with the selection of SPV and PMDC sources, respectively. Similarly, the Q_3 is used during charging and discharging cycles of battery and Q_4 is

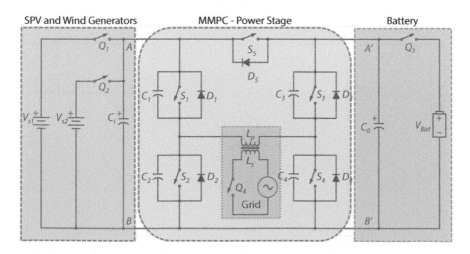

Figure 7.6 Multi-mode power converter topology.

utilized for grid-connection operation. These auxiliary switches can also be replaced by relays and contactors.

The auxiliary switches are used to select the converter modes such as buck, boost, and inverter, whereas the S_1–S_5 are operated in a particular sequence using Digital Pulse Width Modulation (DPWM). The detailed modes of operation of MMPC converter are presented as follows.

7.4.1   Buck Mode

In buck mode, the average output voltage of the converter is always smaller than the input voltage. If the supply voltage, V_{s1} or V_{s2} is greater than the battery voltage, V_{Bat}, and the switch Q_1 or Q_2 is always ON to operate the converter in buck mode.

Applying Kirchhoff's Voltage Law (KVL), when S_1 is ON and D_2 is OFF in Figure 7.7.

$$V_s = V_{Lp} - V_{Co} \tag{7.1}$$

where

V_S is supply or source voltage in Volts.
V_{Lp} is the primary winding voltage of transformer in Volts.
V_{Co} is the voltage across the output filter capacitor in Volts.

$$V_s = L_P \frac{dI_{Lp}}{dT_{on}} - V_{Co} \tag{7.2}$$

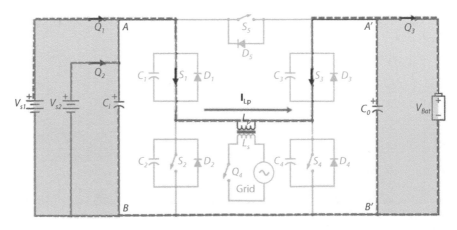

Figure 7.7 MMPC buck mode when switch S_1 is ON.

The inductor voltage depends on the rate of change of current, dI_{Lp}, in the inductor L_p during turn-ON time period, T_{on} of the switch S_1.

$$\frac{dI_{Lp}}{dT_{on}} = \frac{(V_s - V_{Co})}{L_p} \qquad (7.3)$$

Similarly, the output voltage depends on the value output filter capacitor C_o. Thus, the current through the capacitor is given by applying Kirchhoff's Current Law (KCL) in Figure 7.7.

$$I_{Lp} - I_o = C_o \frac{dV_{Co}}{dT_{on}} \qquad (7.4)$$

where

I_{Lp} is the current through the transformer primary winding in Ampere.

I_o is the load current in Ampere.

dV_{Co} is the ripple voltage across the output filter capacitor in Volts.

$$\frac{dV_{Co}}{dT_{on}} = \frac{(I_{Lp} - I_o)}{C_o} \qquad (7.5)$$

Similarly, when S_1 is OFF and D_2 is ON, the supply V_s is isolated from the circuit. Thus, the KVL equation for Figure 7.8 is given by

$$0 = L_p \frac{dI_{Lp}}{dT_{off}} + V_{Co} \qquad (7.6)$$

The rate of change of current in the inductor dI_{Lp} during change in dT_{off} time period is denoted as

$$\frac{dI_{Lp}}{dT_{off}} = \frac{-V_{Co}}{L_p} \qquad (7.7)$$

Similarly, the current through the filter capacitor C_o is derived using KCL

$$I_{Lp} - I_o = C_o \frac{dV_o}{dT_{off}} \qquad (7.8)$$

Thus, the rate of change of voltage dV_{Co} during dT_{off} time period is denoted as

$$\frac{dV_{Co}}{dT_{off}} = \frac{(I_{Lp} - I_o)}{C_o} \qquad (7.9)$$

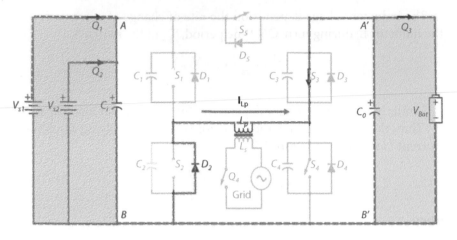

Figure 7.8 MMPC buck mode when switch S$_1$ is OFF.

The voltage balance equation in buck mode of MMPC is given by

$$V_{Co} = V_{Bat} = V_s D \qquad (7.10)$$

where

> V_s is SPV or PMDC voltage in Volts.
> V_{Co} is the voltage across the output filter capacitor C_o in Volts.
> D is duty cycle to the converter, $D = \dfrac{T_{on}}{T}$ and $1 - D = \dfrac{T_{off}}{T}$.
> T is total time of the switching signal, $T = T_{on} + T_{off} = \dfrac{1}{f_s}$.
> T_{on} is turn-ON time period of the switch in Seconds.
> T_{off} is turn-OFF time period of the switch in Seconds.
> f_s is switching frequency in kHz.

7.4.2   Boost Mode

In boost mode of MMPC, the output voltage is always greater than the input source voltage. The proposed converter is operated in bi-directional boost mode. Firstly, if V_{s1} or $V_{s2} < V_{Bat}$, then Q_1 or Q_2 is operated to drive the current from source to get boost output as shown in Figures 7.9 and 7.10.

Applying KVL in Figure 7.9, when S$_1$ is ON and D$_2$ is OFF, the supply voltage is stored in the inductor L$_p$.

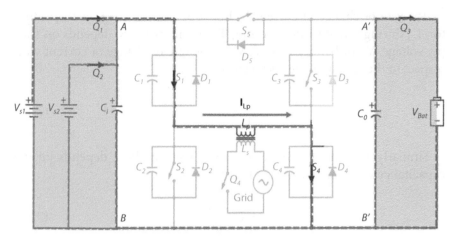

Figure 7.9 MMPC boost mode when switch S$_4$ is ON.

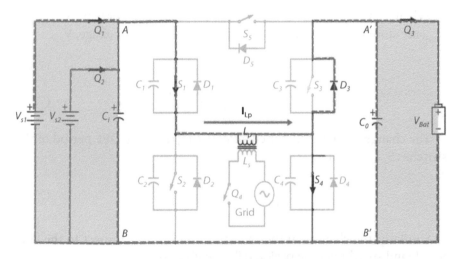

Figure 7.10 MMPC boost mode when switch S$_4$ is OFF.

$$V_s = V_{Lp} \tag{7.11}$$

The inductor voltage, V_{Lp}, depends on the rate of change of current, dI_{Lp}, in the inductor L_p during turn-ON time period, T_{on}, of the switch S_2.

$$V_s = L_p \frac{dI_{Lp}}{dT_{on}} \tag{7.12}$$

The rate of change of current in boost operation, dILp, in the inductor, L_p, during turn-ON time period, T_{on} of the switch S_2 depends on supply voltage, V_s, or battery voltage, V_{Bat}. The rate of change of current with respect to the supply voltage is given by

$$\frac{dI_{Lp}}{dT_{on}} = \frac{V_s}{L_p} \tag{7.13}$$

Similarly, the rate of change of output voltage, dV_{Co}, depends on the capacitance value and its current I_o as follows:

$$I_o = C_o \frac{dV_{Co}}{dT_{on}} \tag{7.14}$$

$$\frac{dV_{Co}}{dT_{on}} = \frac{I_o}{C_o} \tag{7.15}$$

When S_1 is OFF and D_1 is ON, the supply voltage, V_s, or battery voltage, V_{Bat}, are stored in the inductor will influence the output voltage, V_{Co}.

$$V_s = V_{Lp} + V_{Co} \tag{7.16}$$

$$V_s = L_p \frac{dI_{Lp}}{dT_{off}} + V_{Co} \tag{7.17}$$

The changes in the inductor current during the turn-OFF period of the switches S_2 are given by

$$\frac{dI_{Lp}}{dT_{off}} = \frac{V_s - V_{Co}}{L_p} \tag{7.18}$$

Similarly, the changes in the capacitor voltage with respect to the current I_o and switching time period dT_{off} are given by

$$I_{Lp} - I_o = C_o \frac{dV_{Co}}{dT_{off}} \tag{7.19}$$

$$\frac{dV_{Co}}{dT_{off}} = \frac{I_{Lp} - I_o}{Co} \tag{7.20}$$

The voltage balance equation in boost mode of MMPC is given by

$$V_s DT = (V_s - V_{Co})(1 - D)T \tag{7.21}$$

The average output voltage of the boost mode is highly dependent on the turn-OFF time period of the power switches, i.e., $(1-D) = \dfrac{T_{off}}{T}$,

$$V_{Co} = V_{Bat} = \frac{V_s}{(1-D)} \qquad (7.22)$$

The output voltage is controlled by having larger turn OFF and lower turn ON time of the switches.

7.4.3   Bi-Directional Mode

In this mode, the proposed MMPC is operated as a bi-directional mode such as an inverter or rectifier. The main aim of the MMPC is to feed AC power to the grid. In addition, rectifier mode is used to charge the battery when the power generation from SPV and PMDC fails and battery voltage is less than the nominal operating level.

Inversion Mode: As mentioned earlier, MMPC takes power from the solar panel, PMDC generator or battery. In this case, battery power is considered for DC to AC conversion by maintaining Q_3-ON throughout the inversion operation and the switches S_1 and S_4 to get the positive side of an AC output. The switches S_2, S_3 along with S_5 are turned ON to generate the negative half-cycle of an AC output as in Figures 7.11 and 7.12.

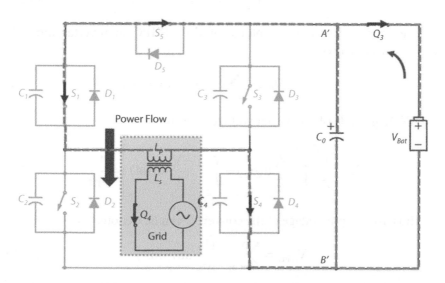

Figure 7.11 MMPC in inverter mode when switches S_1 and S_4 are ON and the remaining are OFF.

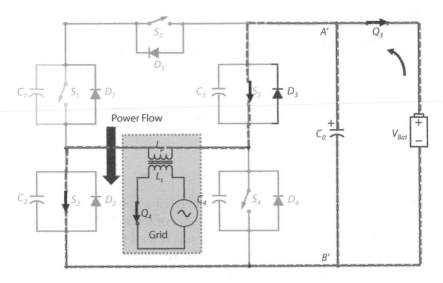

Figure 7.12 MMPC in inverter mode when switches S_5, S_3, and S_2 are ON or inverter.

The output voltage $V_{o,inv}$ and inverter modes are illustrated in Figure 7.11:

$$V_{o,inv} = V_A - V_B \tag{7.23}$$

$$I_{o,inv} = i_A - i_B \tag{7.24}$$

The square wave output voltage of the MMPC in inverter mode is expressed as follows:

$$V_A \text{ or } V_B = \pm \sum_{n=1}^{\infty} V_m \sin(n\omega t) \tag{7.25}$$

The magnitude of the output voltage V_m is obtained as

$$V_m = \frac{1}{\pi}\left[\int_0^{\frac{\pi}{2}} \frac{V_s}{2} d(\omega t) + \int_{\frac{\pi}{2}}^{-\frac{\pi}{2}} \frac{-V_s}{2} d(\omega t)\right] = \frac{4V_{in}}{n\pi} \tag{7.26}$$

Thus, the output voltage of the converter can be denoted as

$$V_{o,inv} = \sum_{n=1}^{\infty} \frac{4V_s}{n\pi} \sin(n\omega t) \tag{7.27}$$

Rectifier Mode: Similarly, during the absence of SPV power and less battery power, the converter will take power from single-phase grid and converted into DC through step-down transformer and diodes in the converter as shown in Figures 7.13 and 7.14. The diodes D_1, D_4, and S_5 are in conduction to rectify positive output and D_2 and D_4 are in conduction to rectify the negative half-cycle of an AC output.

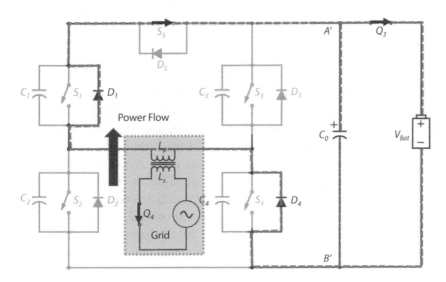

Figure 7.13 MMPC in rectifier mode during D_1, D_4, and S_5 in conduction for positive half-cycle.

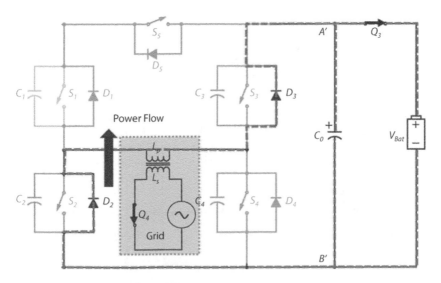

Figure 7.14 MMPC in rectifier mode during D_2 and D_3 in conduction for negative half-cycle.

The output voltage of MMPC in rectifier mode is given as

$$V_{o,rec} = \frac{2 * V_m}{\pi} \tag{7.28}$$

where V_m is the magnitude of the supply voltage to the rectifier in Volts.

7.5 Control Scheme

The continuous monitoring and control of proposed converter in multiple modes are important aspect in smart grid operation. A FPGA controller is best suitable for this application because of its reconfiguration feature. A reconfiguration is a selective updation of a function or the entire FPGA's programmable logic and routing resources while the rest of the device's programmable resources continue to function without interruption. In this work, the low-cost Altera Cyclone IV FPGA is used for mode selection and converter control using digital PWM control.

Figure 7.15 represents the functional block diagram of control logic implemented in FPGA. The built-in clock frequency of FPGA is 50 MHz

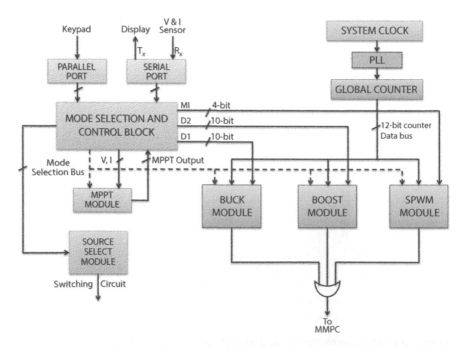

Figure 7.15 FPGA QALU architecture for digital MS-PWM control implementation.

and this is reduced to 10 MHz using Phase-Locked Loop (PLL). This reduced clock frequency from PLL is given to the global counter which will generate a ramp wave at 10 kHz. The pulse width modulated (PWM) signal is generated by comparing a ramp and a reference wave. In Cyclone-IV FPGA, a 10-bit global counter is used to generate a ramp wave. Similarly, the same ramp is compared with a sine wave for SPWM generation to control the inverter operation. This global counter has a zero flag and an overflow flag. The zero flag gets high when the counter reaches 0 and the overflow flag gets high when the counter reaches 999 since the counter counts 1,000 values to generate a ramp wave.

The generated ramp is given to the three modules, namely, buck, boost, and SPWM module. The buck module is used for generating PWM for duty cycle D1. This is given to the switches Q_1 and Q_2. Similarly, the boost module is used for generating PWM for duty cycle D_2. The SPWM module is used for generating suitable modulation index to control inverter operation and this is given to the switches S_1, S_2, S_3, and S_4. All the inputs and control signals for buck boost and SPWM module is given by the mode selection and control block. It is treated as a master control function for the entire operation. It is used to control the operating mode of the circuit depending on the user input. The push buttons are used to select the operating mode manually where 8-bit parallel port is used for the mode selection. The display and measurement unit which consists of a co-processor and LCD module and measuring units is connected to FPGA using a serial Port. The measured value current and voltage sensor from SPV, PMDC, and battery are sent to serial port for mode selection and control.

7.5.1 Mode Selection

The buck, boost, and inverter mode of operation with five sub-modes are considered for the mode selection in MMPC. The mode selection is done based on the voltage level at the source and load terminals. The procedure for the mode selection is illustrated in Figure 7.16.

7.5.2 Maximum Power Point Tracking

This MPPT module gets the voltage and current input from the SPV and PMDC. It runs the MPPT algorithm for tracking maximum power. This output data is given to the control block. The control block decides increment or decrement of the duty cycle based on the input data as illustrated in Figure 7.16. The modified stepped perturb and observe (MSPO) MPPT algorithm as illustrated in Figure 7.17 is implemented using Cyclone IV

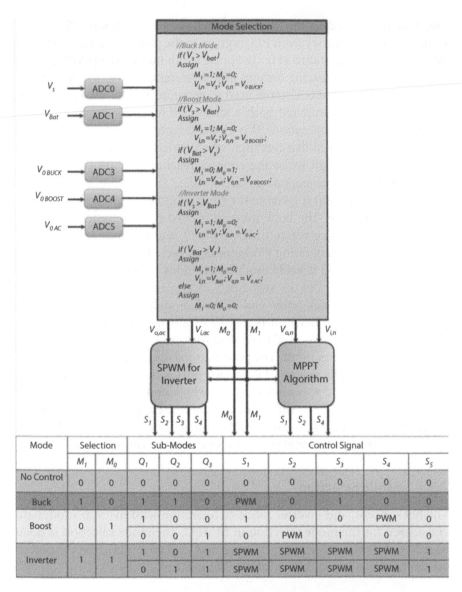

Figure 7.16 FPGA-based mode selection for MMPC.

FPGA. The duty cycle of the PWM pulse is tuned automatically according to the DC input operating point.

If the operating point is far from the maximum value, it increases the step size which enables a fast tracking ability. In buck mode, 12-V, 5-Ah battery is charged using 13.5-V source. Similarly, in boost mode, the output is fixed as 24 V to drive the DC load. Hence, the duty cycle of the

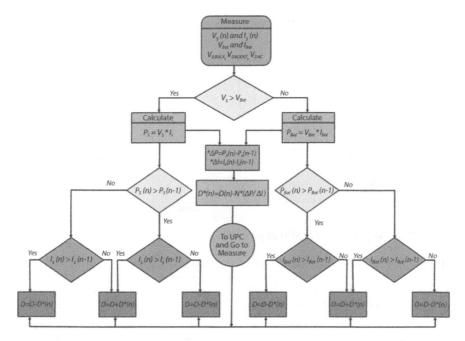

Figure 7.17 Flowchart of MSPO-MPPT algorithm.

converter is varied according to the changes in the input voltage. The proposed FPGA-based MSPO-MPPT controller will adjust the duty cycle of the Buck/Boost converter within a particular range.

7.5.3   Reconfigurable SPWM Generation

The sinusoidal pulse width modulated (SPWM) signal is generated using QALU in Cyclone IV FPGA. The design of single channel SPWM strategy is illustrated in Figure 7.18. The frequency of sawtooth wave is fixed as 10 kHz and it is generated using PLL. The 10-bit counter is enabled to count from 0 to 1023. In this SPWM generation, only positive cycle of 50-Hz sine wave considered comparing with the sawtooth signal for pulse generation. Each comparison count is represented by a time T_1 to T100 and has a value of 100 μS each. The full cycle of sine wave is 20 mS and half cycle is 10 mS. Thus, 100 samples × 100 μS interval = 10 mS data which is presented in Table 7.1.

The output voltage frequency is changed by changing the frequency and voltage magnitude is changed by varying the modulation index data, i.e., predetermined lookup data.

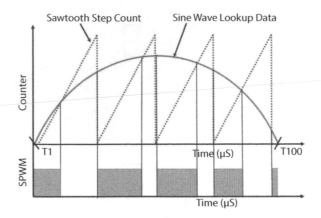

Figure 7.18 Graphical view of SPWM generation.

Table 7.1 Predetermined lookup data for SPWM.

Time (μs)	Data	Time (μs)	Data	Time (μs)	Data	Time (μs)	Data	Time (μs)	Data
T_1	0	T_{21}	604	T_{41}	960	T_{61}	942	T_{81}	574
T_2	33	T_{22}	629	T_{42}	969	T_{62}	931	T_{82}	548
T_3	65	T_{23}	654	T_{43}	976	T_{63}	919	T_{83}	522
T_4	98	T_{24}	678	T_{44}	983	T_{64}	907	T_{84}	496
T_5	130	T_{25}	701	T_{45}	988	T_{65}	893	T_{85}	469
T_6	162	T_{26}	723	T_{46}	993	T_{66}	879	T_{86}	442
T_7	194	T_{27}	745	T_{47}	996	T_{67}	863	T_{87}	414
T_8	226	T_{28}	766	T_{48}	999	T_{68}	847	T_{88}	386
T_9	257	T_{29}	787	T_{49}	999	T_{69}	831	T_{89}	358
T_{10}	288	T_{30}	806	T_{50}	999	T_{70}	813	T_{90}	329
T_{11}	319	T_{31}	824	T_{51}	999	T_{71}	795	T_{91}	301
T_{12}	350	T_{32}	842	T_{52}	999	T_{72}	775	T_{92}	271
T_{13}	380	T_{33}	859	T_{53}	996	T_{73}	756	T_{93}	242
T_{14}	409	T_{34}	875	T_{54}	993	T_{74}	735	T_{94}	212
T_{15}	439	T_{35}	890	T_{55}	988	T_{75}	714	T_{95}	182
T_{16}	468	T_{36}	904	T_{56}	983	T_{76}	692	T_{96}	152

(Continued)

Table 7.1 Predetermined lookup data for SPWM. (*Continued*)

Time (µs)	Data	Time (µs)	Data	Time (µs)	Data	Time (µs)	Data	Time (µs)	Data
T_{17}	496	T_{37}	917	T_{57}	977	T_{77}	670	T_{97}	122
T_{18}	524	T_{38}	929	T_{58}	969	T_{78}	646	T_{98}	92
T_{19}	551	T_{39}	941	T_{59}	961	T_{79}	623	T_{99}	61
T_{20}	577	T_{40}	951	T_{60}	952	T_{80}	598	T_{100}	31

7.6 Results and Discussion

Buck Mode: Figures 7.19 to 7.21 show the MMPC buck mode results with different duty cycles as represented. Voltage developed and current ripples across the inductor are shown by channel 2 and channel 3, respectively. As mentioned in earlier, during buck mode of operation the duty cycle of half-bridge which is immediate follower of source is altered and duty cycle of another half-bridge is always maintained at 100% or always ON state.

Boost Mode: The result obtained from MMPC at boost mode with different duty cycles are presented in Figures 7.22 to 7.25. The voltage developed and current ripples across the inductor are illustrated in channel 3 and channel 4, respectively.

Figure 7.22 shows the result waveforms obtained in boost mode without direct power transfer part. It clearly shows that there is no overlapping time period between first and second duty cycle. Similarly, in Figure 7.23 there is no overlapping but the results is poor due to energy stored in the

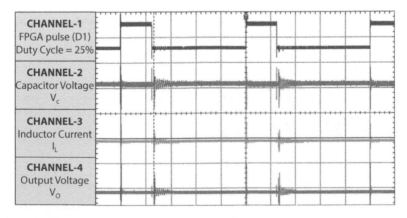

Figure 7.19 Buck mode with duty cycle of 25%.

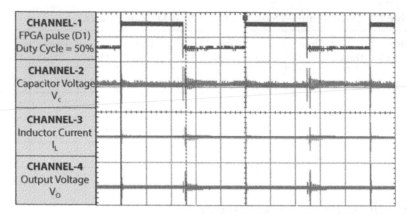

Figure 7.20 Buck mode with duty cycle of 50%.

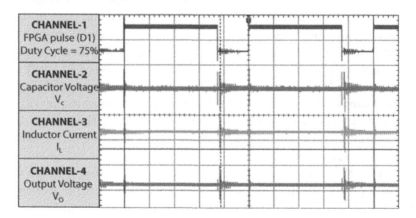

Figure 7.21 Buck mode with duty cycle of 75%.

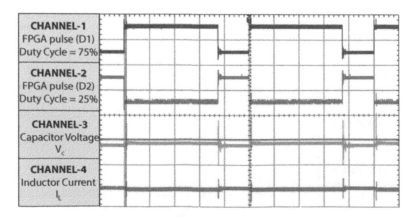

Figure 7.22 Boost mode with duty cycle of 75% and 25%.

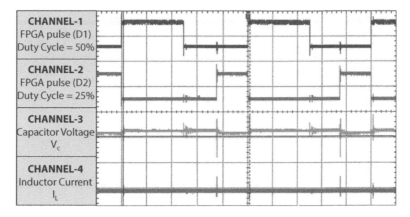

Figure 7.23 Boost mode with duty cycles of 50% and 25%.

inductor. So, both of these conditions should be avoided and this is done by a DC voltage regulator module. In addition, this regulator module ensures that duty cycle falls within the specified limits so that the converter operates at maximum efficiency region.

Figures 7.24 and 7.25 show the result waveforms obtained in boost mode with and without direct power transfer, respectively. The direct power transfer is indicated by overlapping angles. At same load, nearly equal output voltage and current are drawn from the source which is lesser compared with the direct power transfer. Figure 7.25 shows the operating point at the maximum output voltage is obtained at boost mode of operation without compromising efficiency of conversion.

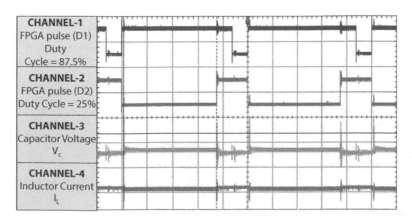

Figure 7.24 Boost mode with duty cycles of 87.5% and 25%.

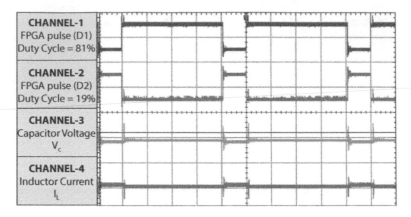

Figure 7.25 Boost mode with duty cycles of 81% and 19%.

Inverter Mode: Figures 7.26 and 7.27 represent the switching pulse generated by FPGA during inverter mode of operation at different modulation indexes.

The modulation index is increased from M = 0.9; the AC voltage magnitude is 219.5 V. The efficiency of the MMPC in the inverter mode is 97% including the switching losses.

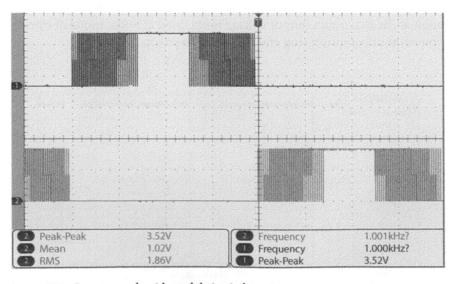

Figure 7.26 Inverter mode with modulation index = 1.

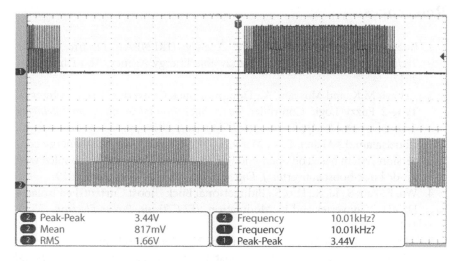

2	Peak-Peak	3.44V		2	Frequency	10.01kHz?
2	Mean	817mV		1	Frequency	10.01kHz?
2	RMS	1.66V		1	Peak-Peak	3.44V

Figure 7.27 Inverter mode with modulation index = 0.9.

7.7   Conclusion

The grid connected MMPC consists of solar PV array and wind energy generator as a primary source of energy and battery as secondary source. The converter system is configured in such a way to operate four-modes. They are buck, boost, inverter and rectifier. The voltage and current sensors are used to monitor the power sources and also the grid line for the synchronized operation. The controller selects the power sources based on the availability of voltage and current at the terminals. The power from solar PV or wind generator is regulated in MMPC to charge the battery and also feeds power to the grid. The various modes of MMPC are presented in detail.

The MS-PWM for a MMPC has been developed and implemented using a Cyclone IV FPGA. The MMPC mode selection and MSPO-MPPT operation are presented for the control of buck/boost converter, and SPWM for inverter operation. The entire digital MS-PWM control is developed using VHDL. The PWM functions such as MSPO and SPWM are stored in the configuration RAM. The accuracy of the functions/algorithms has been improved extensively using resource utilization to the highest level. The proposed FPGA-based controller can be used for power electronics converter applications such as electric vehicles and charging stations.

References

1. International Renewable Energy Agency (IRENEA), *Electric Vehicles: Technology brief.* International Renewable Energy Agency, Abu Dhabi, Feb. 2017.
2. Asy'ari, M.K. and Musyafa, A., Design of Buck Converter Based on Interval Type-2 Fuzzy Logic Controller. *2018 International Seminar on Intelligent Technology and Its Applications (ISITIA),* Aug. 2018, pp. 153–156.
3. Abdelsalam, I., Alajmi, B.N., Marei, M., II, Alhajri, M.F., Wind energy conversion system based on open-end winding three-phase PMSG coupled with ac–dc buck-boost converter. *J. Eng.,* 2019, 17, 4336–4340, April 2019.
4. Wu, F., Fan, S., Li, X., Luo, S., Bidirectional Buck–Boost Current-Fed Isolated DC–DC Converter and Its Modulation. *IEEE Trans. Power Electron.,* 35, 5, 1–11, Oct. 2019.
5. Akhilesh, K. and Lakshminarasamma, N., Control Scheme for Improved Efficiency in a H-bridge Buck-Boost Converter. *2018 IEEE International Conference on Power Electronics, Drives and Energy Systems (PEDES),* Dec. 2018, pp. 1–6.
6. Han, J.-K., Kim, J.-W., Moon, G.-W., A High-Efficiency Asymmetrical Half-Bridge Converter With Integrated Boost Converter in Secondary Rectifier. *IEEE Trans. Power Electron.,* 32, 11, 8237–8242, Nov. 2017.
7. Hassan, M.A., Li, E.-p., Li, X., Li, T., Duan, C., Chi, S., Adaptive Passivity-Based Control of dc–dc Buck Power Converter With Constant Power Load in DC Microgrid Systems. *IEEE J. Emerging Sel. Top. Power Electron.,* 7, 3, 2029–2040, Sept. 2019.
8. Alargt, F.S., Ashur, A.S., Kharaz, A.H., Adaptive delta modulation controller for interleaved buck DC-DC converter. *2017 52nd International Universities Power Engineering Conference (UPEC),* Aug. 2017.
9. Kim, S.-H., Choi, W., Choi, S., Lee, K.-B., Combined Dithered Sigma-Delta Modulation based Random PWM Switching Scheme. *J. Power Electron.,* 9, 5, 667–678, Sept. 2009.
10. Huang, Q., Huang, A.Q., Yu, R., Liu, P., Yu, W., High-Efficiency and High-Density Single-Phase Dual-Mode Cascaded Buck–Boost Multilevel Transformerless PV Inverter With GaN AC Switches. *IEEE Trans. Power Electron.,* 34, 8, 7474–7488, Aug. 2019.
11. Roy, J., Xia, Y., Ayyanar, R., High Step-Up Transformerless Inverter for AC Module Applications With Active Power Decoupling. *IEEE Trans. Ind. Electron.,* 66, 5, 3891–3901, May 2019.
12. Meraj, M., Rahman, S., Iqbal, A., Ben-Brahim, L., Common Mode Voltage Reduction in a Single-Phase Quasi Z-Source Inverter for Transformerless Grid-Connected Solar PV Applications. *IEEE J. Emerging Sel. Top. Power Electron.,* 7, 2, 1352–1363, June 2019.
13. Kumar, A. and Sensarma, P., A Four-Switch Single-Stage Single-Phase Buck–Boost Inverter. *IEEE Trans. Power Electron.,* 32, 7, 5282–5292, July 2017.

14. Wang, H., Han, M., Han, R., Guerrero, J.M., Vasquez, J.C., A Decentralized Current-Sharing Controller Endows Fast Transient Response to Parallel DC–DC Converters. *IEEE Trans. Power Electron.*, 33, 5, 4362–4372, May 2018.

15. Zhang, J., Li, L., Dorrell, D.G., Guo, Y., Modified PI controller with improved steady-state performance and comparison with PR controller on direct matrix converters. *Chin. J. Electr. Eng.*, 5, 1, 53–66, March 2019.

16. Merai, M., Wissem Naouar, M., Slama-Belkhodja, I., Monmasson, E., An Adaptive PI Controller Design for DC-Link Voltage Control of Single-Phase Grid-Connected Converters. *IEEE Trans. Ind. Electron.*, 66, 8, 6241–6249, Aug. 2019.

17. Ahmed, E.M. and Shoyama, M., Variable Step Size Maximum Power Point Tracker Using a Single Variable for Stand-alone Battery Storage PV Systems. *J. Power Electron.*, 11, 2, 218–227, Mar. 2011.

18. Bibin Raj V, S. and Glan Devadhas, G., "Design and development of new control technique for standalone PV system". *Microprocess. Microsyst.*, 72, 1–13, 2020Sept. 2019.

19. Pongiannan, R.K., Sathiyanathan, M., Vinothkumar, U., Mohammed Junaid, K.A., Prakash, A., Yadaiah, N., FPGA–realization of digital PWM controller using Q-format-based signal processing. *J. Vib. Control*, 21, 5, 938–948, July 2013.

20. Sobrino-Manzanares, F. and Garrigos, A., Bidirectional, Interleaved, Multiphase, Multidevice, Soft-Switching, FPGA-Controlled, Buck–Boost Converter With PWM Real-Time Reconfiguration. *IEEE Trans. Power Electron.*, 33, 11, 9710–9721, Nov. 2018.

21. Naouar, W., Ben Hania, B., Slama-Belkhodja, I., Monmasson, E., Naassani, A.A., FPGA-based sliding mode direct control of single phase PWM boost rectifier. *Math. Comput. Simul.*, 91, 1, 249–26, May 2013.

22. Mellit, A., Rezzouk, H., Messai, A., Medjahed, B., FPGA-based real time implementation of MPPT-controller for photovoltaic systems. *Renewable Energy*, 36, 2011, 1652–1661, Dec. 2010.

23. Sadek, U., Sarjaš, A., Svečko, R., Chowdhury, A., "FPGA-based control of a DC-DC boost converter". *IFAC-Papers Online*, 48, 10, 22–27, June 2015.

24. Sathiyanathan, M., Jaganathan, S., Josephine, R.L., Design and Analysis of Universal Power Converter for Hybrid Solar and Thermoelectric Generators. *J. Power Electron.*, 19, 1, 220–233, Jan. 2019.

25. International Solar Energy Map: https://solargis.com/maps-and-gis-data/download/india.

26. Coimbatore Monthly Climate Averages: https://www.worldweatheronline.com/coimbatore-weather-averages/tamil-nadu/in.aspx.

8

Decoupled Control With Constant DC Link Voltage for PV-Fed Single-Phase Grid Connected Systems

C. Maria Jenisha

Department of EEE, National Institute of Technology – Tiruchirappalli,
Tamil Nadu, India

Abstract

A solar energy conversion system feeding to a single-phase utility grid employing SPWM-based dq control has been employed for autonomous control of real and reactive power. The controller consists of a dc-dc boost converter and a voltage source inverter (VSI). In addition to feeding power to the grid, the scheme in this chapter also supplies a local RL load. It is ensured that reactive power of load is supplied from VSI, in order to maintain unity power factor (UPF) at grid terminals. A PI controller is used for varying the duty ratio of the dc-dc converter to maintain the dc link voltage constant. The grid synchronization of VSI is achieved through phase-locked loop (PLL). To evaluate the performance of this scheme, MATLAB/Simulink-based model is tested under varying irradiations. Experiments have been carried out on a PV array of 480 W, 130.8 V, 4.5 A, and the test results are furnished to validate the developed scheme. The controllers used are implemented using dSPACE DS1103 controller.

Keywords: Decoupled control, single phase grid, voltage source inverter, Dspace

8.1 Introduction

Due to the unprecedented development in economy and automation, global claim for electrical energy is increasing day by day. Fossil fuel which

Email: mariajenisha91@gmail.com

M. Kathiresh A. Mahaboob Subahani and G.R. Kanagachidambaresan (eds.) Integration of Renewable Energy Sources with Smart Grid, (171–186) © 2021 Scrivener Publishing LLC

satisfies the foremost part of the energy demand will exhaust in near future, and therefore, it is the necessity of the era to shift for a substitute source for energy. One such substitute source of energy is solar energy, which is a prominent source for electrical energy generation, due to the advances in power electronic conversion techniques, controllers, and its integration. According to the service point of view, solar energy consumption can be classified as

(i)   stand-alone system,
(ii) grid tied system.

Single and two-stage integrations are common in single-phase load/grid connected solar PV system [1–6]. The limitations of the load/grid connected solar PV system lies in its power converters and its control. Sangwongwanich *et al.* and Yao *et al.* have proposed a two-stage PV grid connected system with Hysteresis Current Controller (HCC) in which switching frequency of the inverter is not constant resulting in increased switching losses [7, 8]. In [9, 10], the authors have proposed a HCC with frequency limit control to limit variation in inverter switching frequency which lowers the switching losses. However, the autonomous control of active and reactive power is not feasible with the above-mentioned controllers. In order to achieve autonomous real and reactive power control in single-phase solar PV grid connected system, Hassaine *et al.* proposed a controller with digital pulse width modulation (DPWM) technique and implemented it with PI controller and lookup table. In this constant dc link voltage system, real and reactive power control and grid synchronization are achieved by inverter controller which makes the controller complex [11]. Dhiraj *et al.* have proposed a dq controller for single-phase dynamic voltage restorer (DVR), in which the source considered is a constant dc supply, and hence, there is no scope for MPPT and dc link voltage regulation [12]. Mishra *et al.* have attempted a dq controller for a single stage PV grid connected system to improve voltage profile at the grid terminal. The above controller is complex since it takes care of regulation of voltage at the input of the inverter together with MPPT operation and autonomous control of real and reactive power. The common constraint of single-stage configurations is that the PV array voltage needs to be always higher than the peak value of the grid voltage. Hence, the modulation index of the inverter may not lie between 0 and 1 at all irradiations [13].

The scheme implemented in this chapter consists of two-stage power converter with a dc-dc converter and a single-phase voltage source inverter

(VSI) with decoupled control scheme for controlling the real and reactive power fed to the grid besides supplying an RL load at the grid terminals. A simple PWM control is used for the dc-dc converter for modulating the duty cycle of the gate pulse to maintain the output voltage (dc link voltage) constant. MPPT is also achieved employing a single current sensor at the dc link. When the solar PV generated power is greater than the ac load, the excess power is fed to the grid, and when the PV power is less than the ac load, the deficit power is fed from the grid. The simulation results are validated by comparing with the experimental results. The grid current THD has been brought within the IEEE standards [14] below 5%.

8.2   Schematic of the Grid-Tied Solar PV System

The block schematic of the grid tied solar PV system with dc link voltage controller, MPPT controller, and dq-controller is shown in Figure 8.1.
   The system consists of

   (i)   Solar PV array,
   (ii)   Boost converter,
   (iii) Single-phase VSI.

   The PV array output voltage which is varying with irradiation is maintained at 200 V at the dc link by the boost converter, through a PI controller. The VSI converts the dc link voltage into single-phase ac supply to feed power to the utility grid. In order to achieve independent control of real and reactive power at the inverter terminal, SPWM-based dq controller

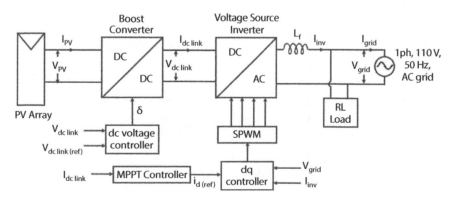

Figure 8.1 Block schematic of the system.

is employed. To extract maximum power at any irradiation, a reference current, $i_{d(ref)}$, is to be determined which is achieved by the MPPT controller. The $i_{d(ref)}$ along with V_{grid} and I_{inv} inputs through dq controller is used to generate the gate pulse for the VSI. Six PV panels rated for 21.7 V open circuit voltage (V_{OC}) each and 4.5 A of short circuit current (I_{SC}) are connected in series to achieve an output power of 480 W which is to be fed to the grid. The power circuit of the grid tied solar PV scheme is shown in Figure 8.2. The experimentally obtained I-V and P-V characteristics of the constructed PV array under varying irradiation are shown in Figure 8.3.

The boost converter output voltage (at the dc link) is given as

$$V_{dc\,link} = \frac{V_{PV}}{1-\delta} \tag{8.1}$$

where δ is the duty ratio.

The inductor and capacitor are chosen so that the ripple current and ripple voltage are within the limits as follows:

$$L = \frac{V_{dc\,link} - V_{PV}(1-\delta)}{\Delta I_{dc\,link} f_s} \tag{8.2}$$

$$C = \frac{I_{dc\,link}\delta}{\Delta V_{dc\,link} f_s} \tag{8.3}$$

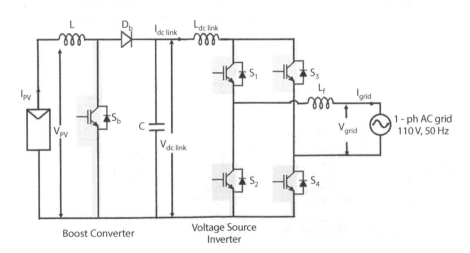

Figure 8.2 Power circuit of the system.

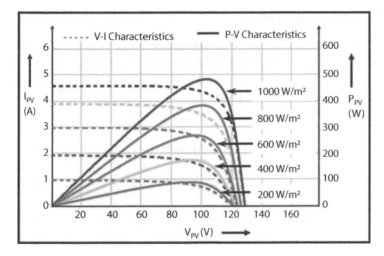

Figure 8.3 Experimentally obtained I-V and P-V characteristics of the PV array.

where $\Delta I_{dc\ link}$ is the ripple current, $\Delta V_{dc\ link}$ is the ripple voltage, and f_s is the switching frequency of the dc-dc converter. Based on Equations (8.3) and (8.4), the critical values of inductor and capacitor are obtained as 7.7 mH and 26 µF, respectively, for $V_{pv} = 96$ V, $V_{dc\ link} = 200$ V, $\delta = 0.5$, and $f_s = 2$ kHz. To operate the converter in continuous conduction mode, the inductor (10 mH) and capacitor (47 µF) values are chosen larger than critical values.

8.2.1 DC Link Voltage Controller

The dc-link voltage is maintained constant at 200 V by automatically adjusting the duty cycle of the boost converter, based on the difference between the actual voltage and the reference voltage. A PI controller and a saw-tooth carrier waveform of 2 kHz are used for generating the gate pulse for the IGBT switch of the boost converter as depicted in Figure 8.4.

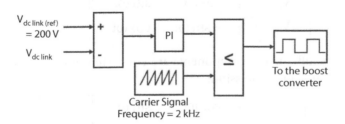

Figure 8.4 Block diagram of the DC link voltage controller.

8.2.2   MPPT Controller

As the voltage output of the boost converter being held constant at 200 V, any change in PV output power reflects as variation in the current flowing in the dc link. This current is sensed and corresponding $i_{d\,(ref)}$ is generated using Perturb and Observe (P&O) technique as expressed in Equations (8.4) and (8.5) for obtaining the maximum power.

$$i_{d(ref)}(k) = i_{d(ref)}(k-1) + \Delta i; \text{ if } i_{dc\,link}(k) > i_{dc\,link}(k-1) \qquad (8.4)$$

$$i_{d(ref)}(k) = i_{d(ref)}(k-1) - \Delta i; \text{ if } i_{dc\,link}(k) < i_{dc\,link}(k-1) \qquad (8.5)$$

where $i_{dc\,link}(k)$ is the present dc link current and $i_{dc\,link}(k-1)$ is the previous instant dc link current. The $i_{d(ref)}$ obtained using the above expression is fed to the dq controller.

8.2.3   SPWM-Based dq Controller

In the SPWM-based dq controller, the existing phase component is considered as α coordinate and the β coordinate is an orthogonal fictitious component to α coordinate. Thus, the generation of αβ components resembles the Clarke's transformation in three phase theory, and consequently, instantaneous αβ values are transformed into dq components by Park transformation. The reference angle (ωt) for Park's transformation is obtained through single-phase PLL.

$$\left. \begin{array}{l} V_\alpha = V_{grid} \sin\omega t \\ V_\beta = V_{grid} \sin(\omega t - 90°) \end{array} \right\} \qquad (8.6)$$

The β coordinate generation has been implemented with the quarter cycle time delay. The αβ voltage components are converted into dq0 components using the following expression:

$$\begin{bmatrix} V_d \\ V_q \end{bmatrix} = \begin{bmatrix} \cos\omega t & \sin\omega t \\ -\sin\omega t & \cos\omega t \end{bmatrix} \begin{bmatrix} V_\alpha \\ V_\beta \end{bmatrix} \qquad (8.7)$$

Similarly, αβ current components are converted into dq0 components using the following expression:

$$\begin{bmatrix} i_d \\ i_q \end{bmatrix} = \begin{bmatrix} \cos\omega t & \sin\omega t \\ -\sin\omega t & \cos\omega t \end{bmatrix} \begin{bmatrix} i_\alpha \\ i_\beta \end{bmatrix} \qquad (8.8)$$

The control logic consists of d-axis and q-axis current controllers, which will independently control the active and reactive power, respectively, as shown in Figure 8.5. The reference currents and measured currents are compared, the outputs of which are regulated by proportional plus integral (PI) controller. The d-axis reference current ($i_{d\,(ref)}$) corresponding to the required real power to be fed to the grid is computed by MPPT controller. In order to pump power to the grid at unit power factor (UPF), the reactive power required by the load is supplied by the inverter. To achieve this, the required value of q-axis reference current ($i_{q\,(ref)}$) is given as input to the controller.

The dq control equations are given as follows:

$$u_d = v_d^* - \omega i_q L_f + v_d \tag{8.9}$$

$$u_q = v_q^* - \omega i_d L_f + v_q \tag{8.10}$$

in which, $\omega = 2\pi f$

$$v_d^* = \left(k_p + \frac{k_i}{S}\right)(i_{d\,ref} - i_d) \tag{8.11}$$

$$v_q^* = \left(k_p + \frac{k_i}{S}\right)(i_{q\,ref} - i_q) \tag{8.12}$$

The dq components are transformed into αβ components as follows:

$$u_\alpha = u_d \cos \omega t - u_q \sin \omega t \tag{8.13}$$

$$u_\beta = u_d \cos \omega t - u_q \sin \omega t \tag{8.14}$$

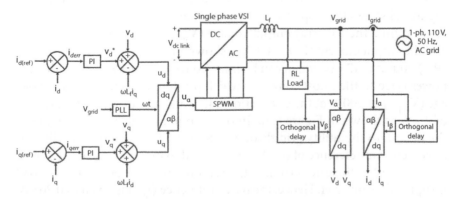

Figure 8.5 Switching technique for SPWM-based VSI controller.

The u_α is the modulating signal that is compared with carrier signal to perform SPWM, and the resulting pulses are given to the appropriate switches of the VSI, while the u_β component is discarded.

8.3   Simulation and Experimental Results of the Grid Tied Solar PV System

To demonstrate the performance of the developed system, the simulation study was conducted in MATLAB/Simulink software prior to experimental investigation. The investigational arrangement of the solar PV single-phase grid connected system in this chapter involves six PV panels connected in series yielding PV array voltage (V_{OC}) of 130 V and PV array current (I_{SC}) of 4.5 A. The PV array output voltage is fed to a boost converter to obtain a steady dc voltage of 200 V. The boost converter is constructed employing IGBT switch (CT60AM: 900 V, 60A) together with gate driver circuit (HCPL 3101). The VSI which has been constructed using IGBT (SKM150GB12T4: 1,200 V, 150 A) switches receives the boost converter output and feeds the corresponding power to the grid. Inductors of 20 and 40 mH are used at the dc link and at the grid terminal, respectively, to appropriately condition the inrush current and grid current to meet IEEE 519 grid standards [14]. Three separate controllers are applied in the present system—one for sustaining the output of the boost converter at 200 V, the other one for obtaining maximum power from PV array, and the third one for the de-coupled control of the VSI. All three controllers are implemented using dSPACE DS1103 system. The required voltages and currents for the dSPACE controller are measured through LV 25-P and LA 55-P transducers.

The experiment has been carried out for different irradiations and Figure 8.4 shows the response of the system discussed for an irradiation of 1,000 W/m². The PV array output voltage (V_{PV}) at this irradiation is 110 V and the output current (I_{PV}) is 4.3 A as shown in Figure 8.4a. The PV power (P_{PV}) output is 473 W which can be inferred from Figure 8.3. Now, this PV power is fed to the Boost converter which produces a fixed dc output voltage $(V_{dc\,link})$ of 200 V at the duty ratio of 45% and an output current $(I_{dc\,link})$ of 2.2 A as shown in Figure 8.4b. In order to pump the power available at the dc link to the utility grid through VSI, the MPPT controller will produce a current reference of $(i_{d\,ref})$ of 2.97 A and to maintain the unity power factor (UPF) at the grid terminal, the required reactive power by the load is supplied by inverter. Hence, the q axis reference $(i_{q\,ref})$ is given as 0.868 A. Since a RL load of 220 W and 67.5 VAr is connected at the inverter output

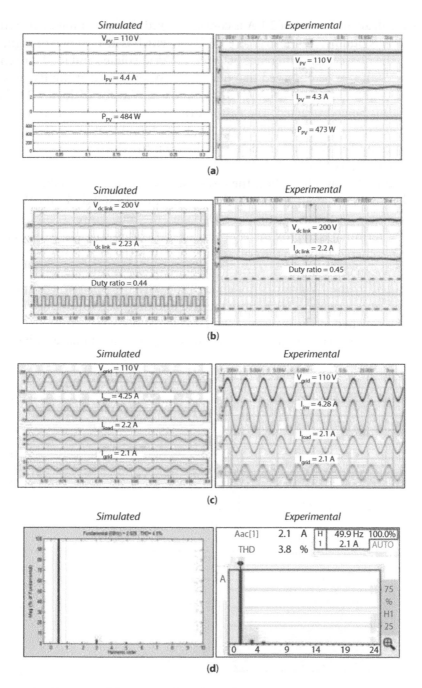

Figure 8.4 Response of the system when the irradiation is at 1,000 W/m²; voltage and current (a) at the PV array terminal, (b) at the dc link terminal, (c) at the grid terminal, and (d) THD of the grid current.

terminal, an ac current of 2.1 A is consumed and the remaining ac current of 2.1 A is pumped to the grid at UPF (I_{grid}), which is also shown in Figure 8.4c. The phase angle difference between the grid voltage and load current is 17.75°. It can be observed that out of 473 W produced at the PV array output, a total power of 449.2 W (220 W and 67.5 VAr for RL load + 228 W to grid) is being extracted. The Fast Fourier Transform (FFT) analysis has been carried out for the grid current and it is found that Total Harmonic Distortion (THD) is 3.8% which is well within the IEEE 519 standards as shown in Figure 8.4d.

Figures 8.5a and 8.5b show the response of the discussed configuration at the inverter output terminal when irradiation is 800 and 600 W/m², respectively. At the irradiation of 800 W/m², the total power produced at the PV terminal is approximately equal to the load power and there is no power exchange between the grid and the inverter. The $i_{d\,ref}$ is 2.97 A and $i_{q\,ref}$ is 0.686 A in this condition. At the irradiation of 600 W/m², the total power produced by the PV array is not enough to satisfy the load power and so the deficit real power is taken from the grid and the reactive power requirement is satisfied by the inverter. When the deficit power is consumed from the grid by the load, the grid current and voltage are out of phase with each

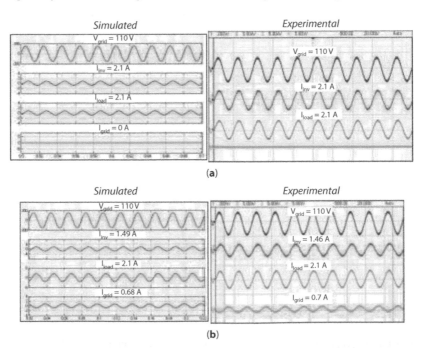

Figure 8.5 Voltage and current at the output of the inverter terminal, when irradiation is (a) 800 W/m² and (b) 600 W/m².

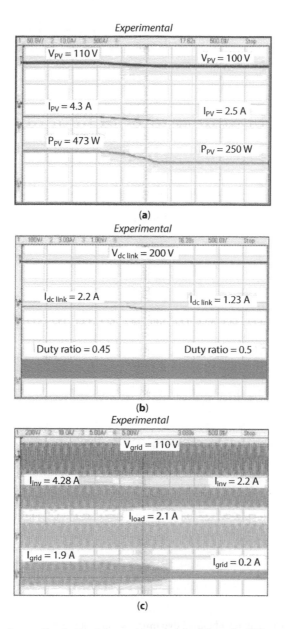

Figure 8.6 Experimentally obtained dynamic response of the solar PV system when irradiation changes from 1,000 to 850 W/m² (a) at the PV terminal, (b) at the dc link, and (c) at the grid terminal.

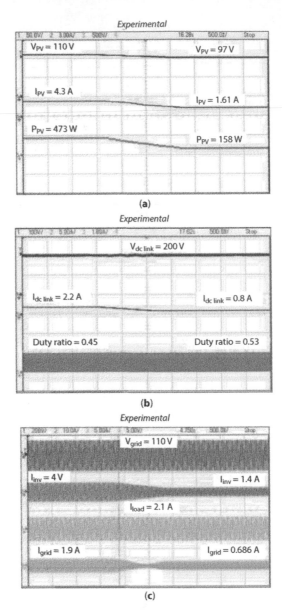

Figure 8.7 Experimentally obtained dynamic response of the projected system when irradiation changes from 1,000 W/m² to 600 W/m² (a) at the PV terminal, (b) at the dc link, and (c) at the grid terminal.

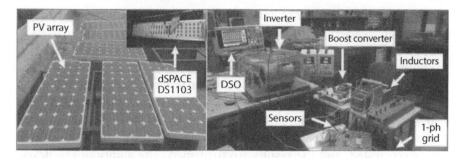

Figure 8.8 Photograph of the experimental setup.

other because of the reversal of current direction as shown in Figure 8.5a. In order to extract maximum power to the grid, the $i_{d\,ref}$ generated by the MPPT controller is 2.1 A and $i_{q\,ref}$ given is 0.686 A.

To study the dynamic behavior of the suggested system, the irradiation is changed from 1,000 W/m² to 850 W/m². The change in irradiation reflects in both PV array output voltage and current; correspondingly, the power changes from 473 to 250 W as shown in Figure 8.6a. The current at the output of the boost converter changes from 2.2 to 1.23 A, $V_{dc\,link}$ being held constant at 200 V as in Figure 8.6b. The inverter output terminal current changes from 4.28 to 2.1 A; correspondingly, the grid terminal current changes from 1.9 to 0.2 A after supplying the load requirement of 2.1 A with the grid voltage of 110 V as shown in Figure 8.6c.

The demonstrated configuration is also subjected to the irradiation change from 1,000 to 600 W/m². As the AC load is 220 W and 69 VAr, at 1,000 W/m², irradiation PV supplies both grid and load, but as soon as the irradiance level changes to 600 W/m², PV generated power is not sufficient to supply the load. Hence, the deficit power is consumed from the grid as shown in Figure 8.7.

It is found that at any irradiation of the PV array, the critical load is satisfied by the PV power or by the grid power. The photograph of the experimental setup of solar PV grid connected system and the accompanying components is shown in Figure 8.8.

8.4   Conclusion

A simple two-stage power electronic interface using boost converter and VSI with decoupled control for grid tied PV array system has been modeled and tested. Experiments have been conducted on a PV array of 480 W and a single-phase utility grid of 110 V, 50 Hz. The I-V and P-V characteristics of the constructed PV array have been obtained under different irradiations through experiments, and the change in irradiation for the analysis of the projected scheme has been replicated using variable dc source based on the above characteristics. A dq controller has been developed for the VSI using which autonomous control of both real and reactive power fed to the grid is achieved with MPPT. The dc link voltage is maintained constant at 200 V in order to make sure the modulation index of the VSI is less than 1 at all irradiations. The steady state and dynamic response of the system have been tested, and the results are furnished. The ability of the developed system to control both the real and the reactive power fed to the grid has been demonstrated using simulation and experimental investigations by employing additional RL load at the inverter output terminal. The gate pulses of IGBTs for both boost converter and VSI are provided from dSPACE RTI 1103 controller. The quality of the grid current is maintained to satisfy IEEE standard and the grid synchronization is achieved using PLL. It is believed that such PV-fed single-phase systems with independent control of real and reactive power would find more domestic applications.

References

1. Chen, Y. and Smedley, K.M., A cost-effective single-stage inverter with maximum power point tracking. *IEEE Trans. Power Electron.*, 19, 1289–1294, 2004.
2. Jain, S. and Agarwal, V., A Single-Stage Grid Connected Inverter Topology for Solar PV Systems With Maximum Power Point Tracking. *IEEE Trans. Power Electron.*, 22, 1928–1940, 2007.
3. Gonzalez, R., Gubia, E., Lopez, J., Marroyo, L., Transformerless Single-Phase Multilevel-Based Photovoltaic Inverter. *IEEE Trans. Ind. Electron.*, 55, 2694–2702, 2008.
4. Araujo, S.V., Zacharias, P., Mallwitz, R., Highly Efficient Single-Phase Transformerless Inverters for Grid-Connected Photovoltaic Systems. *IEEE Trans. Ind. Electron.*, 57, 3118–3128, 2010.
5. Xiao, H., Xie, S., Chen, Y., Huang, R., An Optimized Transformerless Photovoltaic Grid-Connected Inverter. *IEEE Trans. Ind. Electron.*, 58, 1887–1895, 2011.
6. Jeylani, M.A., Mahaboob Subahani, A., Kanakaraj, J., Rameshkumar, K., Analysis of Solar PV Application based on Bidirectional Inverter. *Proc. META Web of Conferences*, September 2018, vol. 225, pp. 1–6, 2018.
7. Sangwongwanich, A., Yang, Y., Blaabjerg, F., A Sensorless Power Reserve Control Strategy for Two-Stage Grid-Connected PV Systems. *IEEE Trans. Power Electron.*, 32, 11, 8559–8569, 2017.

8. Yao, Z. and Xiao, L., Two-Switch Dual-Buck Grid-Connected Inverter with Hysteresis Current Control. *IEEE Trans. Power Electron.*, 27, 7, 3310–3318, 2012.

9. Ho., C.N., Cheung., V.S.P., Chung., H.S., Constant-Frequency Hysteresis Current Control of Grid-Connected VSI Without Bandwidth Control. *IEEE Trans. Power Electron.*, 24, 11, 2484–2495, 2009.

10. Wu., F., Zhang., L., Wu., Q., Simple unipolar maximum switching frequency limited hysteresis current control for grid-connected inverter. *IET Power Electron.*, 7, 4, 933–945, 2014.

11. Hassaine., L., Olias, E., Quintero., J., Haddadi., M., Digital power factor control and reactive power regulation for grid-connected photovoltaic inverter. Elsevier - *Renewable Energy*, 34, 1, 315–321, 2009.

12. Katole, D.N., Daigavane, M.B., Gawande, S.P., Daigavane, P.M., Modified Single Phase SRF d-q Theory Based Controller for DVR Mitigating Voltage Sag in Case of Nonlinear Load. *J. Electromagn. Anal. Appl.*, 9, 22–33, 2017.

13. Mishra., S. and Mishra., Y., Decoupled controller for single-phase grid connected rooftop PV systems to improve voltage profile in residential distribution systems. *IET Renewable Power Gener.*, 11, 2, 370–377, 2017.

14. *IEEE Std 519-2014 (Revision of IEEE Std 519-1992) - IEEE Recommended Practice and Requirements for Harmonic Control in Electric Power Systems.*

9

Wind Energy Conversion System Feeding Remote Microgrid

K. Arthishri[1]* and N. Kumaresan[2]

[1]School of Electrical and Electronics Engineering, SASTRA Deemed to be University, Thanjavur, Tamilnadu, India
[2]Department of Electrical and Electronics Engineering, National Institute of Technology Thiruchirappalli, Tiruchirappalli, Tamilnadu, India

Abstract

Every aspect of electrical power generation, transmission, distribution, and utilization, including the associated controls, has been improving in performance, owing to the steady advancement in technology, over the decades. In urban areas, three-phase power has been made readily available for industrial loads at higher voltages such as 11 kV and 6.6 kV and for domestic needs at 400 V. However, there are many remote/rural locations that are yet to be powered due to the lack of accessibility to the mainland and the electric grid. In such locations, considering the technical challenges and economic considerations, the trend is toward establishing a single-phase grid in the preliminary stage, which can be upgraded later to a three-phase system, when the load requirement increases substantially. In this context, the present approach is to make use of wind or solar energy or both as a hybrid system. This chapter deals with conventional and modern configurations along with their closed-loop control strategies for feeding power to remote microgrids, using wind energy electric conversion systems.

For wind energy conversion system (WECS) of rating more than 3 kW, it is preferable to employ a three-phase induction generator and operate it for delivering single-phase output for feeding microgrids.

While configuring a three-phase generator for single-phase output, care should be taken to reduce the unbalance in the three phase stator currents of the generator. Further, it is to be noted that if the generators are directly connected to the grid, the rotor speed variation from no load to full load is very small, which restricts the peak power extraction from the wind source.

*Corresponding author: arthishriranga@gmail.com

M. Kathiresh A. Mahaboob Subahani and G.R. Kanagachidambaresan (eds.) Integration of Renewable Energy Sources with Smart Grid, (187–208) © 2021 Scrivener Publishing LLC

Keywords: Electric machine analysis computing, induction generators (IGs), peak power tracking, steady-state analysis, wind power generation

9.1   Introduction

Access to reliable energy is an important requirement for enhancing the quality of life and health for the people living in rural/remote areas. Uninterrupted power supply is also very much needed for providing water supply for agricultural operations, as it is the main occupation of more than 50% of the population in our country. So, ensuring the economic growth of this population is a must to sustain the overall development of any nation [1]. However, to extend the power grid to all these locations is presently not viable because of huge investment and technical difficulties in enlarging the transmission line system for this purpose [2]. Further, in the recent years, installation of distributed generators in the distribution networks has been intensified to effectively utilize the renewable resources available at the consumer end, in addition to taking steps to reduce the losses in the transmission network [3, 4]. Some of the typical units across the globe for such application are as follows: (i) Buru Island in Indonesia: According to the Indonesian power survey 2018, it is proposed that at least 25% of power generation in Indonesia should be met by renewable energy in 2025 [5]; (ii) Isle of Wight Island in UK: According to the report from the department of Civil, Environment and Geothermal Engineering, UK, a 100% renewable energy system is planned to be implemented in this island [6]; (iii) Andaman Island in India: It is currently powered by diesel generators and hydro plant. It is also planned to supply at least 25% of the demand through renewable energy sources [7]. Considering these facts and the falling cost of installation of renewable energy systems, employing wind-driven generators appears to be a viable option for electrifying rural and remote locations [8, 9].

As stated earlier, in view of the locational limitations, it is cost-effective to install single-phase grid in the preliminary stage, by employing three-phase generators, but operating them to deliver single-phase power output. In choosing the generator to be coupled with the wind turbine (WT) for power generation, several options are available such as squirrel-cage IGs (IGs), wound rotor IGs and permanent magnet synchronous generators [10, 11]. Among these, Squirrel-Cage Induction Generators (SCIG) are widely preferred due to their less maintenance, light weight, ease of availability, rugged construction absence of slip rings and DC excitation, and low cost. So, SCIGs have been largely employed as Wind-Driven Induction Generators (WDIG) for both grid-connected and stand-alone modes of operation [12–16]. In this chapter, the grid connected mode of operation of the WDIG is discussed in detail.

9.2   Literature Review

Owing to technological improvements and significant cost reductions, power generation from wind energy is nearly competitive with conventional fuels. The cumulative installed capacity of wind power installations has reached close to 650 GW in 2019 as per the report of Global Wind Energy Council shown in Figure 9.1 [17]. So, wind energy electric conversion system has come to play a vital role in the power scenario and with varying wind speed; the system should be designed as a maximum power point tracking (MPPT) model.

In the early stages of development, WDIGs were directly connected to the three-phase grid. In this configuration, the stator frequency of the generator is same as the grid frequency and the rotor speed has a narrow range of operation, resulting in limiting the maximum power extraction from the wind [13–16]. In [10], De Mello and Harmnett have presented the complete review on the application and the performance of

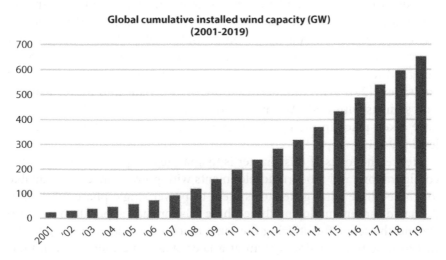

Figure 9.1 Cumulative installed capacity of wind.

IGs for wind energy conversion. Squirrel-cage induction machines were employed in commercial wind farms worldwide, since, apart from the other advantages mentioned earlier, it does not also require any synchronising equipment for integrating with the grid. In such grid connected IGs, while delivering real power to the grid, the machines draw reactive power from the grid itself.

Subsequently, it was found that WDIGs with smaller power ratings can be conveniently employed for supplying isolated loads, thereby reducing the grid dependence. In such systems, the reactive power requirement of these generators can be locally met by operating them as Self-Excited Induction Generators (SEIGs), by capacitors across the stator terminals [18–20]. With continued research in the analysis, design and operation of SEIGs for supplying standalone loads, the power rating of the generators has steadily increased, and now, they are proposed for forming microgrids [21–23]. The future trend would be running several SEIGs in parallel to increase the rating of these grids [24].

The reactive power control in SEIGs has also been described in the literature using star-delta stator winding switching schemes [14, 15]. The detailed investigations have also been carried out on the effect of voltage profile and steady-state voltage stability margin in radial distribution system with the addition of WDIG. Later, to incorporate the variable speed operation of the turbine for peak power extraction, various power electronic circuit configurations and control schemes have been developed for WDIGs [25–28]. Senthil Kumar *et al.* have interfaced the WDIG using a diode bridge rectifier and a voltage source converter to the three-phase grid [25]. Here again, the induction machine is operated as SEIG and the power fed to the microgrid has been controlled with an analog-based hysteresis current controller (HCC). Also, the three-phase SEIG has been designed for feeding DC microgrid using appropriate power electronic converters [23]. With the increased use of DC loads and power generation sources with both AC and DC outputs, considerable developments have taken place in the hybrid operation of AC/DC at the distribution level [29–31].

Hence, the focus of this chapter is to analyze and present conventional configurations and modern developments with power converters, suitable for three-phase wind generator system to power single-phase microgrid. As stated earlier, employing three-phase IGs for such application will also expand the scope for utilizing the same machine, while upgrading to the three-phase transmission system at a later stage. The merits and limitations of each configuration are also discussed on the basis of desired performance quantities of the system.

9.3   Direct Grid Connected Configurations of Three-Phase WDIG Feeding Single-Phase Grid

The direct grid connected configuration of three-phase WDIG system for feeding a single-phase grid is shown in Figure 9.2. The major drawback with this system is the unbalance in the three-phase stator currents, as only two phases of the stator are connected to the grid. This also leads to the derating of the generator.

With a view to minimize this unbalance, the three-phase stator winding is configured in various ways with phase balancing capacitors (PBCs) as shown in Figure 9.3 for feeding power to the single-phase grid [32–34]. In Figures 9.3a and b, single and two PBCs are used to minimize the current unbalance. However, here the unbalance cannot be eliminated and hence the problem of derating of the IG still exits. In Smith connection shown in Figure 9.3c, the current unbalance is reduced only for a particular operating condition but it is still unable to completely nullify the unbalance for the entire operating range. Another limitation of these configurations is the narrow range of operating speed of the system that restricts maximum power extraction from the wind source.

To overcome these shortcomings, instead of PBCs, suitable power converters are interfaced between the wind-generator system and the single-phase grid.

9.4   Three-Phase WDIG Feeding Single-Phase Grid With Power Converters

The overall setup of the wind energy conversion system, where the three-phase variable voltage–variable frequency (VV-VF) output of the generator is connected to the single-phase microgrid through power converters is shown in Figure 9.4. Here, the WT is coupled to the SCIG with a capacitor bank at the stator terminals of the generator.

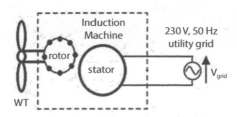

Figure 9.2 Direct grid connected configuration of three-phase WDIG feeding single-phase grid.

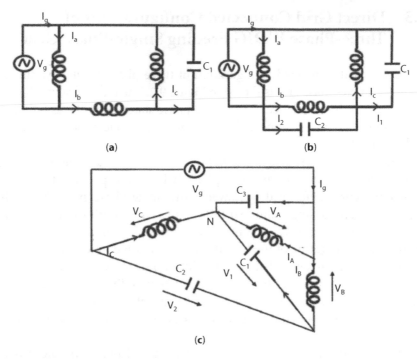

Figure 9.3 Stator winding configurations for three-phase induction generator supplying single-phase grid (a) Single capacitor configuration; (b) Two capacitor configuration; (c) Smith configuration.

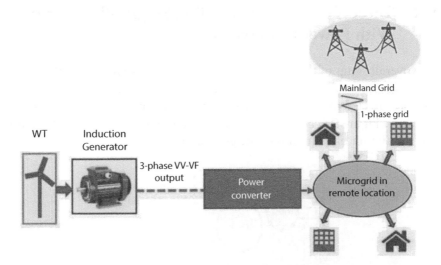

Figure 9.4 Overall setup of the three-phase wind generator system feeding power to single-phase grid.

Following are the various power converter configurations for the operation of three-phase WDIGs with single-phase grid.

1. Uncontrolled Rectifier–Line Commutated Inverter (UR-LCI)
2. Uncontrolled Rectifier–DC/DC Converter–Line Commutated Inverter [UR-(DC/DC)-LCI]
3. Uncontrolled Rectifier–Voltage Source Inverter (UR-VSI)

The significant features of these configurations along with the control strategies are illustrated and compared. A systematic procedure to determine the end-to-end performance of the wind generator system is shown in Figure 9.4 is also given in the following sections. As the WT and SEIG remain common for all the configurations, the performance predetermination procedure at maximum power condition for these machines is given in the succeeding sections.

9.5 Performance of the Three-Phase Wind Generator System Feeding Power to Single-Phase Grid

The performance of the system shown in Figure 9.4 is estimated in three stages, namely, WT characteristics, SEIG performance estimation, and power converter performance, as shown in Figure 9.5.

9.5.1 Wind Turbine Characteristics

In the first stage, the values of the rotor speed (N_{rotor}) and the maximum mechanical power input (P_M) to the generator (i.e., output of WT) are

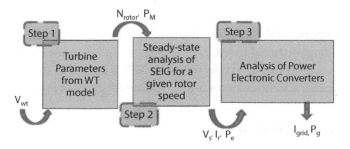

Figure 9.5 Step-by-step procedure for the performance evaluation of wind energy conversion system.

evaluated from the WT characteristics at which the maximum power can be extracted. These values can be obtained using the well-known mathematical model of WT.

The mechanical output power of the WT is given by [35–39]

$$P_M = 0.5 C_P(\lambda, \beta) \rho \alpha A V_{wt}^3 \tag{9.1}$$

where
  λ - tip speed ratio,
  β - pitch angle in degrees
  ρ_a - air density in kg/m³
  A - swept area in m²
  V_{wt} - wind velocity in m/s
$C_p(\lambda, \beta)$ - power coefficient

The power coefficient is given by

$$C_P(\lambda, \beta) = c_1(c_2/\lambda_i - c_3\beta - c_4)e^{-c_5/\lambda_i} + c_6\lambda \tag{9.2}$$

where $\dfrac{1}{\lambda_i} = \dfrac{1}{\lambda + 0.08\beta} - \dfrac{0.035}{\beta^3 + 1}$

The c_1 to c_6 are coefficients whose values depend upon the type of WT. The tip speed ratio is given by

$$\lambda = R\omega_{rT}/V_{wt} \tag{9.3}$$

where ω_{rT} is the rotational speed of WT in rad/s and R is the radius of the blade in m.

The mechanical torque produced by the WT can be expressed as

$$T_m = P_M/\omega_{rT} \tag{9.4}$$

The power curve of the WT, i.e., the mechanical output power, P_M, and the corresponding rotational speed, N_{rotor} ($=2\pi\omega_{rT}$) can be drawn using the above expressions for various wind velocities. The MPPT curve can then be obtained and typical characteristics for a 4.5-kW turbine are given in Figure 9.6.

9.5.2   Generator Analysis

In the case of generators directly connected to a power grid, the terminal voltage and frequency are known, and hence, their analysis is

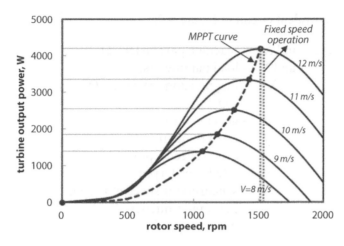

Figure 9.6 Wind turbine output power curve for various wind velocities.

straightforward. However, for a SEIG with power converter interface, both terminal voltage and frequency are unknown and they also vary with the prime-mover speed and load impedance. Further, owing to the saturation, the magnetizing reactance varies with the operating point. Hence, it is essential to arrive at a modified equivalent circuit incorporating the effect of VV-VF to predetermine the performance of such SEIGs.

Such modified equivalent circuit is shown in Figure 9.7 and the following assumptions are made:

(i)   Only the magnetizing reactance is considered to vary with the level of saturation, all other parameters being constant.
(ii) The core loss in the machine (R_m) is neglected.
(iii) All the parameters are referred to stator side.
(iv) Reactances are given at rated frequency.

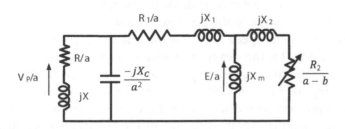

Figure 9.7 Modified equivalent circuit of SEIG.

In the equivalent circuit of Figure 9.7,

R_1, R_2, X_1, and X_2 - machine parameters, Ω
R and X - load parameters, Ω
X_m - magnetizing reactance, Ω
s - slip
f_g - generated frequency, Hz
f_r - rated frequency, Hz
N - actual speed, rpm
N_r - rated speed, rpm
N_s - synchronous speed at f_g, rpm
a - per unit frequency
b - per unit speed

Then,

$$a = f_g / f_r \tag{9.5}$$

$$b = N/N_r \tag{9.6}$$

$$N_r = 120 f_r/P \tag{9.7}$$

$$N_s = 120 f_g/P = a120 f_r/P \tag{9.8}$$
$$\text{Slip (s)} = (N_s - N)/N_s$$

$$\therefore S = \frac{\dfrac{a120 f_r}{P} - \dfrac{b120 f_r}{P}}{\dfrac{a120 f_r}{P}} = \left[\frac{a-b}{a}\right] \tag{9.9}$$

With this modified equivalent circuit, the performance estimation is carried out by the following procedure.

For a given wind velocity, the estimation of the steady-state performance of WDIG can be started with the rotational speed and the maximum mechanical power input to the generator attained from Step 1 as indicated in Figure 9.6. Having speed (N_{rotor}) as the known parameter, the modified equivalent circuit for the steady-state analysis of SEIGs is shown in Figure 9.7 is to be solved to calculate the generator performance. In this circuit, for any operating condition, the value of machine parameters, load parameters and excitation capacitive reactance (X_C) are known. But for any given b, a and X_m are unknown. So, the first step in the analysis is to evaluate these unknown parameters. For calculating the generator performance, effective value of load at the generator terminals $(R/a + jX)$ is required. The angle between the phase voltage and fundamental component of the current at the generator terminals is assumed to be zero in all the configurations discussed here. So, the load in the equivalent circuit of Figure 9.7 is considered to be resistive (R_e) and for any given generator phase voltage (V_p) and the power output (P_e) from the generator, the value of this resistance can be written as

$$R_e = 3V_P^2/P_e \qquad (9.10)$$

So, for a given pu speed, by knowing the value of R_e from (9.10), the value of excitation capacitance and machine parameters of the SEIG, the unknown parameters a and X_m can be evaluated using binary search algorithm as presented in [40]. In this evaluation, if X_m calculated is more than the critical value of X_m (X_{mc}), then the excitation of SEIG fails and ceases its operation [41]. If X_m is less than X_{mc}, the induced emf (E/a) can be calculated using the experimentally obtained magnetization characteristics of the test machine.

For obtaining the magnetization characteristic of the induction machine, (i) the rotor of the induction machine was driven by a separately excited DC motor at constant speed, and (ii) stator was supplied with the rated frequency. Under this condition, the rotor side circuit given in Figure 9.7 will be open circuited, and hence, the magnetizing current, I_m, will flow through $(R_1 + j\ X_1)$ and X_m only. At each voltage setting, the applied voltage, V_{in}, and the corresponding magnetizing current, I_m, were measured.

Then, X_m and the corresponding air-gap voltage (E/a) at each setting can be calculated as

$$X_m = \sqrt{(V_{in}/I_m)^2 - R_1^2} - X_1 \qquad (9.11)$$

$$(E/a) = X_m I_m \qquad (9.12)$$

where V_{in} is the per phase input voltage and I_m is the per phase input current of the machine when operated with $a = b$.

Having evaluated E/a, other performance quantities of SEIG can be calculated using the following expressions:

$$\text{Stator voltage, } V_P = E\frac{Z_{LC}}{Z_S + Z_{LC}} \qquad (9.13)$$

$$\text{Load current, } I_P = \frac{V_P/a}{Z_L} \qquad (9.14)$$

$$\text{Capacitor current, } I_C = \frac{V_P/a}{Z_C} \tag{9.15}$$

$$\text{Stator current, } I_S = I_C + I_P \tag{9.16}$$

$$\text{Rotor current, } I_R = \frac{E/a}{Z_R} \tag{9.17}$$

$$\text{Electrical power output, } P_e = 3\text{re}\left(V_P I_P^*\right) \tag{9.18}$$

$$\text{Stator copper loss, } P_S = 3\times|I_S|^2\times R_1 \tag{9.19}$$

$$\text{Rotor copper loss, } P_R = 3\times|I_R|^2\times R_2 \tag{9.20}$$

$$\text{Mechanical power input, } P_M = 3|I_R|^2\ R_2\left(\frac{b}{a-b}\right) \tag{9.21}$$

$$= P_e + P_S + P_R$$

In order to arrive at the peak power point of the WT, the equivalent resistance (R_e) given in (9.10) has to be tuned to match the P_{mp} for a given wind velocity. The step-by-step process involved in such tuning is given in [42]. The outputs V_P, I_P and P_e thus obtained from this process can be used as the input for the analysis of power converter at step 3.

9.6   Power Converter Configurations

The various power converter configurations for three-phase WDIG feeding power to single-phase grid and their performance quantities along with the closed-loop operation are discussed in the following subsections.

9.6.1   Configuration 1: WDIG With Uncontrolled Rectifier–Line Commutated Inverter

Figure 9.8 shows the UR-LCI power converter configuration in which the three-phase variable AC output of the WDIG is converted to DC using an UR. A LCI is operated in a closed-loop, by adjusting the firing angle (α) to transfer the maximum possible power from the wind-generator system to the single-phase grid. In this configuration, a sufficiently large value of

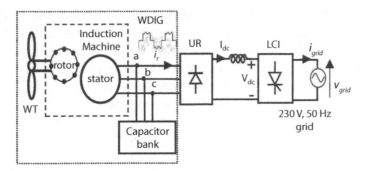

Figure 9.8 UR-LCI configuration.

inductor is needed at the DC side of the UR to maintain the DC current constant and continuous. Under this condition, the performance quantities at the generator load terminals are as follows:

$$I_{r1} = \sqrt{6}I_{dc}/\pi \tag{9.22}$$

$$I_r = \sqrt{2/3}I_{dc} \tag{9.23}$$

$$\text{Power Factor} = 3/\pi \tag{9.24}$$

$$\text{THD} = \sqrt{I_r^2 - I_{r1}^2}/I_{r1} \tag{9.25}$$

where I_{dc} is the average value of current at the DC side of the three-phase UR as well as at the input side of the single-phase LCI.

Using (9.22)–(9.25), THD of the generator load current with three-phase UR can be calculated as 31.08%. Similarly, the performance quantities of the single-phase LCI are as follows:

$$V_{dc} = -2\sqrt{2}V_{grid}\cos\alpha/\pi \tag{9.26}$$

$$I_{grid1} = 2\sqrt{2}I_{dc}/\pi \tag{9.27}$$

$$I_{grid} = I_{dc} \tag{9.28}$$

$$P_g = V_{grid}I_{grid1}\cos\alpha \tag{9.29}$$

$$\text{Power factor} = (I_{grid1}/I_{grid})\cos\alpha \tag{9.30}$$

$$\text{THD} = \sqrt{I_{grid}^2 - I_{grid1}^2}/I_{grid1} \tag{9.31}$$

Using (9.31), the THD of the grid current can be evaluated as 48.34% with constant and continuous I_{dc}. From (9.26)–(9.30), it is evident that the power delivered to the grid and the power factor predominantly depends on the firing angle α. To achieve very near Unity Power Factor (UPF) operation, the firing angle of LCI should be maintained close to 180° [43–45]. In this case, the power fed to the grid will not be the maximum possible value. So, the reduced power extraction from the wind source and the reactive power burden on the grid due to the low power factor operation of LCI makes this system not very suitable to operate the three-phase wind-driven IG with the single-phase grid. Apart from this, passive tuned LC filters are required to achieve sinusoidal current at the grid terminal. Inclusion of such passive filters will bring down the overall efficiency of the system due to the power loss in these filter components.

9.6.2   Configuration 2: WDIG With Uncontrolled Rectifier–(DC-DC)–Line Commutated Inverter

To overcome the drawback of UR-LCI configuration such as low power factor, a configuration with an additional DC-DC converter shown in Figure 9.9 can be employed. In the case of LCI, if it is less than 90°, there will be a power flow from the grid to the dc-link and if the firing angle is greater than 90°, the power is fed from the dc-link to the grid. Also, the dc-link voltage should be greater than the gird voltage.

9.6.2.1   Closed-Loop Operation of UR-DC/DC-LCI Configuration

In this configuration, the DC-DC converter can be operated for peak power point tracking in the closed loop as shown in Figure 9.10. The LCI

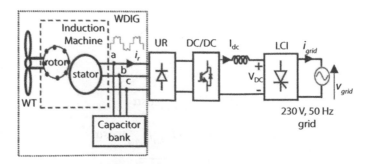

Figure 9.9 UR-(DC-DC)-LCI configuration.

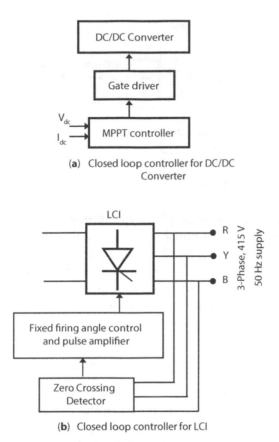

(a) Closed loop controller for DC/DC Converter

(b) Closed loop controller for LCI

Figure 9.10 Closed-loop operation of WDIG-DC/DC–LCI configuration.

is operated in closed loop with constant firing angle (α) close to 180°. In fact, such a configuration has been proposed in the literature for supplying three-phase grid in wind energy conversion system [43–45]. This configuration also requires filters for mitigating the THD of the current fed to the grid terminal. Having an additional power conversion stage is another shortfall of the system shown in Figure 9.9.

9.6.3   Configuration 3: WDIG With Uncontrolled Rectifier–Voltage Source Inverter

Figure 9.11 shows the UR-VSI configuration for the operation of three-phase WDIG with single-phase grid. In this system also, the three-phase AC output from the generator is converted into a DC using the UR. This

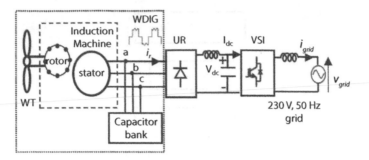

Figure 9.11 UR-VSI configuration.

DC is converted into a single-phase AC through a VSI tied to the grid. In this configuration, it is possible to achieve UPF at the grid with peak power extraction from the wind source by controlling the IGBTs of the VSI. It is important to maintain the grid current sinusoidal with limited filter requirement in the real-time applications. Also, the possibility of maintaining UPF at the grid terminals makes the UR-VSI a better alternative as compared to the LCI configurations.

9.6.3.1   Closed-Loop Operation of UR-VSI Configuration

For the closed-loop operation of the system shown in Figure 9.11, the VSI is operated to track the peak power of the WT and to maintain UPF at the grid. In order to extract maximum power from the wind, there exists a critical operating speed of WT. In the present system, output current of the inverter is monitored and varied appropriately so that WT is operated at the specific speed. This is possible because of the fixed grid voltage and UPF operation of the inverter [42, 46]. In such a case, the rise or fall in the wind power will correspondingly reflect at the power fed to the grid by changing the grid current. Therefore, a closed-loop control technique is required for generating the reference grid current (i_{ref}) to extract the maximum power available in the wind, and Figure 9.12 shows the overall control scheme.

Firstly, the dc link voltage and the current are sensed and multiplied to calculate the dc power. This value of dc power is used to generate the peak value of reference current using the perturb and observe algorithm. The peak value of the current thus generated is multiplied with the unit sine wave generated from the grid voltage (v_{grid}). The sinusoidal quantity thus obtained is compared with the actual grid current (i_{grid}), and the error is fed to the HCC. This controller generates the gate pulses for triggering the IGBTs in such a way that the current fed to the grid stays within the

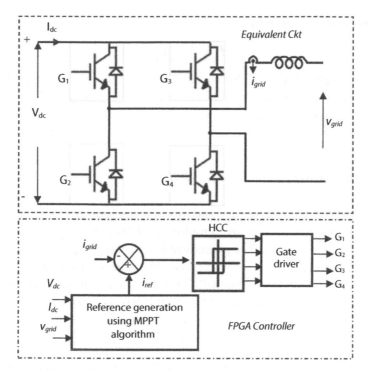

Figure 9.12 Closed-loop operation of UR-VSI configuration.

upper and the lower bands of the reference current, yielding UPF at the grid terminals.

The output voltage of the UR shown in Figure 9.12 can be written as

$$V_r = \left(3\sqrt{2}V_L/\pi\right) \tag{9.32}$$

V_L is the line voltage at the input of the rectifier which is same as V_p in the present case.

The DC voltage and current at the inverter terminal can be written as

$$V_{dc} = V_r - I_{dc} \times R_L \tag{9.33}$$

$$I_{dc} = \left(\pi/\sqrt{6}\right)I_{RL} \tag{9.34}$$

where R_L is the resistance of the inductor and the fundamental component of the input current to the rectifier (I_{RL}) is written as

$$I_{RL} = \sqrt{3}I_P \tag{9.35}$$

The grid power is given by

$$P_g = V_{grid}I_{grid} \tag{9.36}$$

However, the limitation of this configuration is that the current at the rectifier input is not sinusoidal and hence contains harmonics. To overcome this, UR can be replaced with a controlled rectifier, thereby maintaining the rectifier input current sinusoidal in the closed-loop operation.

9.7 Summary

In this chapter, the conventional and modern configurations of three-phase wind energy conversion system employing SCIG feeding power to single-phase grid are discussed. The procedure involved in the performance analysis of SEIG is explained in detail. Also, the closed-loop operation of the power converter configurations with LCI and VSI is presented. In addition, a schematic procedure for the end-to-end performance evaluation right from the WT to the grid is illustrated. To summarize, for a three-phase WDIG feeding remote grid (single-phase).

- Conventional configurations have limitations of stator current unbalance, derating of the IG, poor power factor, and restricted power output from the wind.
- With the modern power converter configurations with LCI, the current unbalance can be eliminated. However, complex filter requirements and low power factor at the grid side are the major limitations.
- The power converter configuration with VSI facilitates balanced stator current, maximum power extraction from the wind source, simple filter design, and UPF at the grid side.
- The UR at the generator terminals in the UR-VSI configuration results in non-sinusoidal current at the generator terminals that results in high Total Harmonic Distortion. However, it is possible to obtain sinusoidal current by using a controlled rectifier at the generator terminals.

References

1. United Nations, Division for Sustainable Development Department of Economic and Social Affairs United Nations, *A Survey of International Activities in Rural Energy Access and Electrification*, 2014, May, [Online]. Available.

2. Hansen, L.H., Helle, L., Blaabjerg, F., Ritchie, E., MunkNielsen, S., Bindner, H., Sorensen, P., Bak-Jensen, B., *Conceptual survey of Generators and Power Electronics for Wind Turbines*, Riso National Laboratory, Roskilde, Denmark, Dec. 2001.

3. Marei, M., II, El-Saadany, E.F., Salama, M.M.A., A Flexible DG Interface Based on a New RLS Algorithm for Power Quality Improvement. *IEEE Syst. J.*, 6, 1, 68–75, Mar. 2012.

4. Xia, N., Gooi, H.B., Chen, S., Hu, W., Decentralized State Estimation for Hybrid AC/DC Microgrids. *IEEE Syst. J.*, 12, 1, 434–443, Mar. 2018.

5. PwC and APLSI, *Alternating Currents: Indonesian Power Industry Survey 2018*, 2018.

6. Farooq, O., *A 100 % Renewable Energy System for Isle of Wight*, University College London, 2014, Retrieved from https://www.cege.ucl.ac.uk/teaching/MSc_Posters /..MSc../ 070_FarooqO_ESE.pdf.

7. Government of India, *Power for All*, 2016, Retrieved from https://powermin.nic.in/sites/default/files/uploads/joint_initiative_of_govt_of_india_and_andman_nicobar.pdf.

8. Nema, P., Nema, R., Rangnekar, S., A current and future state of art development of hybrid energy system using wind and PV-solar: A review. *Renewable Sustainable Energy Rev.*, 13, 8, 2096–2103, 2009.

9. Mandelli, S., Barbieri, J., Mereu, R., Colombo, E., Off-grid systems for rural electrification in developing countries: Definitions, classification and a comprehensive literature review. *Renewable Sustainable Energy Rev.*, 58, 1621–1646, May 2016.

10. De Mello, F.P. and Harmnett, L.N., Large Scale Induction Generators for Power Systems. *IEEE Trans. Power App. Syst.*, PAS-100, 5, 2610–2618, May 1981.

11. Shadhu Khan, K. and Chatterjee, J.K., Three-Phase Induction Generators: A Discussion on Performance. *Electr. Mach. Power Syst.*, 27, 8, 813–832, 1999.

12. Bansal, R.C., Bhatti, T.S., Kothari, D.P., Bibliography on the application of induction generators in nonconventional energy systems. *IEEE Trans. Energy Convers.*, 18, 3, 433–439, Sep. 2003.

13. Kumaresan, N. and Subbiah, M., Innovative Reactive Power Saving in Wind-driven Grid-connected Induction Generators Using a Delta-Star Stator Winding: Part II, Estimation of Annual Wh and VARh of the Delta-Star Generator and Comparison with Alternative Schemes. *Wind Eng.*, 27, 3, 195–204, May 2003.

14. Raja, P., Kumaresan, N., Subbiah, M., Grid-Connected Induction Generators Using Delta-Star Switching of the Stator Winding with a Permanently Connected Capacitor. *Wind Eng.*, 36, 2, 219–231, Apr. 2012.

15. Raja, P., Selvan, M., Kumaresan, N., Enhancement of voltage stability margin in radial distribution system with squirrel cage induction generator based distributed generators. *IET Gener. Transm. Distrib.*, 7, 8, 898–906, Jul. 2013.

16. Boldea, I., *Variable Speed Generators*, Second Edition, CRC Press, Taylor & Francis Group, US, 2015.

17. Global Wind Energy Council (GWEC), *Global Wind Statistics*, 2017, Retrieved from https://gwec.net/wp-content/uploads/vip/GWEC_PRstats2017_EN-003_FINAL.pdf.
18. Singh, G.K., 'Self-excited induction generator research- A survey. *Electr. Power Syst. Res.*, 69, 4, 107–114, 2004.
19. Bansal, R.C., Bhatti, T.S., Kothari, D.P., Discussion of Bibliography on the Application of Induction Generators in Nonconventional Energy Systems. *Energy*, 19, 3, 832463, 2004.
20. Simoes, M.G. and Felix, A.F., *Modeling and Analysis with Induction Generators*, CRC Press, Taylor & Francis Group, US, 2014.
21. Krishnan, A., Natarajan, K., Dwivedi, A.C., Energy Saving Scheme for Grid Connected Pico-Hydro Energy Conversion System Supplying Hybrid Loads. *2019 National Power Electronics Conference (NPEC)*, Tiruchirappalli, India, pp. 1–5, 2019.
22. Kumaresan, N., Analysis and control of three-phase self-excited induction generators supplying single-phase AC and DC loads. *IEE Proc. Electr. Power Appl.*, 152, 39–747, 2005.
23. Mahaboob Subahani, A., Senthil Kumar, S., Kumaresan, N., Real and reactive power control of SEIG systems for supplying isolated DC loads. *J. Inst. Eng. India Ser. B*, 99, 6, 587–595, 2018.
24. Essaki Raj, R., Kamalakannan, C., Karthigaivel, R., Genetic algorithm-based analysis of wind-driven parallel operated self-excited induction generators supplying isolated loads. *IET Renewable Power Gener.*, 12, 4, 472–483, 2018.
25. Senthil Kumar, S., Subbiah, M., Kumaresan, N., Analysis and control of capacitor-excited induction generators connected to a micro-grid through power electronic converters. *IET Gener. Transm. Distrib.*, 9, 10, 911–920, Jun. 2015.
26. Bueno, E., Cobreces, S., Rodriguez, F., Hernandez, A., Espinosa, F., Design of a Back-to-Back NPC Converter Interface for Wind Turbines with Squirrel-Cage Induction Generator. *IEEE Trans. Energy Convers.*, 23, 3, 932–945, Sep. 2008.
27. Pucci, M. and Cirrincione, M., Neural MPPT Control of Wind Generators with Induction Machines Without Speed Sensors. *IEEE Trans. Ind. Electron.*, 58, 1, 37–47, Jan. 2011.
28. Chen, Z., Guerrero, J., Blaabjerg, F., A Review of the State of the Art of Power Electronics for Wind Turbines. *IEEE Trans. Power Electron.*, 24, 8, 1859–1875, Aug. 2009.
29. Das, D., Kumaresan, N., Nayanar, V., Navin Sam, K., Ammasai Gounden, N., Development of BLDC Motor-Based Elevator System Suitable for DC Microgrid. *IEEE/ASME Trans. Mechatron.*, 21, 3, 1552–1560, Jun. 2016.
30. Strunz, K., Abbasi, E., Huu, D.N., DC Microgrid for Wind and Solar Power Integration. *IEEE J. Emerging Sel. Top. Power Electron.*, 2, 1, 115–126, Mar. 2014.

31. Tan, K.T., Sivaneasan, B., Peng, X.Y., So, P.L., Control and Operation of a DC Grid-Based Wind Power Generation System in a Microgrid. *IEEE Trans. Energy Convers.*, 31, 2, 496–505, Jun. 2016.
32. Radwan, A.A. and Mohamed, Y.A.R., II, Networked Control and Power Management of AC/DC Hybrid Microgrids. *IEEE Syst. J.*, 11, 3, 1662–1673, Sep. 2017.
33. Chan, T., Performance analysis of a three-phase induction generator connected to a single-phase power system. *IEEE Trans. Energy Convers.*, 13, 3, 205–213, Sep. 1998.
34. Chan, T. and Lai, L.L., Single-phase operation of a three-phase induction generator with the Smith connection. *IEEE Trans. Energy Convers.*, 17, 1, 47–54, Mar. 2002.
35. Wang, Q. and Chang, L., An Intelligent Maximum Power Extraction Algorithm for Inverter-Based Variable SpeedWind Turbine Systems. *IEEE Trans. Power Electron.*, 19, 5, 1242–1249, 2004.
36. Muyeen, S.M. and Al-Durra, A., Modeling and control strategies of fuzzy logic controlled inverter system for grid interconnected variable speed wind generator. *IEEE Syst. J.*, 7, 4, 817–824, 2013.
37. Vijayakumar, K., Tennakoon, S.B., Kumaresan, N., Ammasai Gounden, N.G., Real and reactive power control of hybrid excited wind-driven grid-connected doubly fed induction generators. *IET Power Electron.*, 6, 6, 1197–1208, 2013.
38. Ghosh, S. and Senroy, N., Electromechanical Dynamics of Controlled Variable-Speed Wind Turbines. *IEEE Syst. J.*, 9, 2, 639–646, 2015.
39. Koutroulis, E. and Kalaitzakis, K., Design of a maximum power tracking system for wind-energy-conversion applications. *IEEE Trans. Ind. Electron.*, 53, 2, 486–494, 2006.
40. Arthishri, K., Anusha, K., Kumaresan, N., Senthil Kumar, S., Simplified methods for the analysis of self-excited induction generators. *IET Electr. Power Appl.*, 11, 9, 1636–1644, 2017.
41. Karthigaivel, R., Kumaresan, N., Subbiah, M., Analysis and control of self-excited induction generator-converter systems for battery charging applications. *IET Electr. Power Appl.*, 5, 2, 247–257, Feb. 2011.
42. Arthishri, K., Kumaresan, N., Ammasai Gounden, N., Analysis and Application of Three-Phase SEIG With Power Converters for Supplying Single-Phase Grid from Wind Energy. *IEEE Syst. J.*, 13, 2, 2018.
43. Binu Ben Jose, D.R., Ammasai Gounden, N.G., Rajesh, V., Power electronic interface with maximum power point tracking using line-commutated inverter for grid-connected permanent magnet synchronous generator. *Electr. Power Compon. Syst.*, 43, 5, 543–555, 2015.
44. Binu Ben Jose, D.R., Ammasai Gounden, N., Ravishankar, J., Simple power electronic controller for photovoltaic fed grid-tied systems using line commutated inverter with fixed firing angle. *IET Power Electron.*, 7, 6, 1424–1434, 2014.

45. Binu Ben Jose, D.R., Vasanth, A.G., Ammasai Gounden, N.G., Hybrid power electronic controller for combined operation of constant power and maximum power point tracking for single-phase grid-tied photovoltaic systems. *IET Power Electron.*, 7, 12, 3007–3016, 2014.
46. Arthishri, K., Kumaresan, N., Ammasaigounden, N., Analysis and MPPT control of a wind-driven three-phase induction generator feeding single-phase utility grid. *J. Eng.*, 6, 220–231, 2017.

10

Microgrid Protection

Suman M.[1]*, Srividhya S.[2] and Padmagirisan P.[3]

[1]Department of EE, MNNIT Allahabad, Prayagraj, India
[2]Department of EEE, NIT Tiruchirappalli, Tiruchirappalli, India
[3]University of Surrey, Guildford, United Kingdom

Abstract

The technological advancements paved a way to deploy energy resources in the distribution system. When these energy resources and loads are operated in parallel, it is cumulatively called as a microgrid. The microgrid offers numerous pros, *viz.,* reduced carbon emission, increased reliability, free of cost fuel, and reduced losses. Despite numerous advantages, it brings up several challenges in the power system perspective. One such prominent challenge is the microgrid protection. The microgrid protection is different from the conventional protection. It is due to the following reasons, a) current is no more unidirectional, b) topology may vary according to the necessity, c) change in fault current due to the varying nature of energy resources input in the distribution system, d) insignificant fault current contribution in the islanded system, and e) insignificant fault current difference due to the shorter lengths. Due to these reasons, protecting microgrid with the existing over-current relay seems to be an impractical. Existing overcurrent relays may fail to detect the fault due to the reduction in the fault current. This chapter elaborates the literature with their pros and cons of various methods.

*Keywords***:** Microgrid, protection, unplanned island formation, overcurrent protection, distance protection, differential protection

10.1   Introduction

The conventional power system comprises centralized power generation, transmission, and distribution. The power generation stations are located at

------

Corresponding author: sumanmnitt@gmail.com

M. Kathiresh A. Mahaboob Subahani and G.R. Kanagachidambaresan (eds.) Integration of Renewable Energy Sources with Smart Grid, (209–238) © 2021 Scrivener Publishing LLC

remote places according to the availability of resources or easy transportation of the resources and transmit the generated power through the transmission system. The power transmission is done at higher voltage level, owing to the advantage of reduced power losses since the current magnitude is reduced, and distributes the power to the customers through the distribution system. The voltage level is reduced at the distribution system depending on the industrial or residential customers. The conventional power system has the following demerits: the conventional energy resources like coal are diminishing, after certain years, there may be no coal to generate power at the thermal power station and the technical losses due to long distance power transmission, etc. Hence, this conventional power system is facing a dramatic change, owing to the development of technologies.

10.2   Necessity of Distributed Energy Resources

As mentioned in the introduction, increase in demand leads to install/ develop new transmission line infrastructure which is extremely higher in cost. Transmitting power over a large distance increases the power loss the conventional energy resources like coal, gasoline, and nuclear are ever diminishing. In order to overcome the aforementioned drawbacks, there is a necessity to generate power nearer to consumer end in order to avoid the transmission losses. Also, if the distributed power generation is based on renewable energy resources, the fuels are available at free of cost. Instead of the conventional centralized power generation, distributed energy resources are available as a distributed manner. DERs are of lower capacity and would be much closer to the distribution network also. The applications of the DERs are prime-power, back-up power, peak shaving, low cost energy, and combined heat and power. The main advantages of distributed energy resources are reduction in transmission and distribution losses, reduces the overloading of transmission system, improves reliability, delays or avoids the new infrastructure of the transmission system, environmentally friendly fuels, and fuels are free of cost if the renewable energy is used [1–3].

10.3   Concept of Microgrid

Microgrids are having multiple definitions provided by various research groups.

The US Department of Energy (DOE) defines the microgrid as follows [4]:

"A microgrid is a group of interconnected loads and distributed energy resources within clearly defined electrical boundaries that act as a single controllable entity with respect to the grid. A microgrid can connect and disconnect from the grid to enable it to operate in both grid-connected or island-mode. A remote microgrid is a variation of a microgrid that operates in islanded conditions."

On the other hand, the International Council on Large Electrical Systems (CIGRE) defined the microgrid in the following way [5]:

"Microgrids are electricity distribution systems containing loads and distributed energy resources (such as distributed generators, storage devices, or controllable loads) that can be operated in a controlled, coordinated way either while connected to the main power network or while islanded."

The microgrid is the replica of the macro-grid, which comprises everything as in the utility grid. Microgrid plays a vital role in smart grid. Microgrids can be operated in two modes: grid connected mode (GCM) and islanded mode. In GCM, the power is exchanged between the microgrid and macro-grid depending on the requirement. However, in islanded mode, the loads are energized only by the DERs exists in the distribution network.

The microgrid offers several advantages such as improved efficiency, reduction in generation cost, power quality improvement, improves reliability, and promotes resiliency. Even though microgrid offers several benefits, it brings up several challenges, one such is protection of microgrids [1–5].

10.4   Why the Protection With Microgrid is Different From the Conventional Distribution System Protection

In general, overcurrent relays are mostly sufficient for the conventional distribution system protection because of the following reasons [6–10]:

    a.   Power flow is unidirectional (from the substation to the load).
    b.   System topology is fixed.
    c.   Only the grid-connected mode of operation is possible since there are no distributed energy resources available.

However, in microgrid, power flow is bi-directional, system topology varies to improve the performance, and islanded mode of operation of the distribution network is possible. These are a few of the several reasons why the protection of microgrid is different from conventional.

10.4.1   Role of the Type of DER on Protection

The DERs are broadly classified into two types based on how it is interfaced with the grid [11]:

(i)   Converter interfaced
(ii)   Rotating machine interfaced

The aforementioned types of interfacing offer different characteristics in protection and modes of operation.

i) Converter Interfaced Distributed Generation (CIDG)
In general, most of the DER modules are CIDG, e.g., PV, fuel cell, and PMSG. In this type, the source may be direct current (DC)/alternating current (AC).

However, irrespective of the type of sources (DC or AC), it is coupled to the grid through a voltage source converter (VSC). The DER modules may be dispatchable or non-dispatchable (operated at MPPT). The CIDGs supply different fault currents depend on their controls (voltage, current, or power control). Irrespective of the type of control, the fault current can be a maximum of 2 p.u., above which the converter damages or converter protection operates from supplying more/further current.

ii) Rotating machine interfaced Distributed Generation (RIDG)
On the other hand, RIDG is interfaced with the grid through machine terminals directly, e.g., squirrel cage induction generator, diesel generator, doubly-fed induction generator, and gas turbine generator (low speed). The RIDG contributes higher amount of fault current.

10.5   Foremost Challenges in Microgrid Protection

There are few challenges which should be carefully considered in developing an effective microgrid protection. The possible challenges are given as follows [6–10].

10.5.1   Relay Blinding

When the microgrid operates in grid-connected mode, if the DG exists between the utility grid and the fault as shown in Figure 10.1, then the

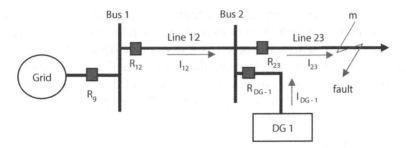

Figure 10.1 Blinding of relays.

fault current supplied by the utility grid may be reduced to a certain extent. Since, the DGs are partly supplying to the fault, resulting in non-detection of faults by the relay (R_{12}) or delayed tripping or non-tripping of the relays, if inverse over current relays are utilized.

10.5.2 Variations in Fault Current Level

In the system shown in Figure 10.2, if a fault is occurred at the location "m", for this fault, the fault current is contributed by the grid (I_{fg}), DG-1 (I_{fDG-1}) and DG-2 (I_{fDG-2}). So, the overall fault current can be expressed as follows:

$$I_f = I_{fg} + I_{fDG-1} + I_{fDG-2}$$

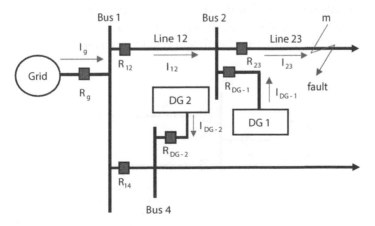

Figure 10.2 Variations in fault current level.

The presence of DGs increased the fault current contribution considerably. The increased fault current may affect the relay-relay, fuse-fuse, and fuse-relay coordination. The protection devices are designed based on the short circuit capacity. The introduction of DGs and the contribution of the DGs during the fault surpasses the capacity of the protection devices, which will eventually damage the protection devices permanently.

10.5.3   Selectivity

Selectivity in protection generally refers to selecting and isolating only the faulty of the system. It is a major concern in microgrid, even though selectivity may be comparatively easier in radial system, it is more complicated in the mesh systems as well as if the microgrid is supplied by a greater number of DGs at various locations of the system.

10.5.4   False/Unnecessary Tripping

Some of the relays in the system may trip falsely due to the inclusion of the DGs. For example, if a fault is occurred at location "m" in the system as shown in Figure 10.3, it is sufficient for the relay R_{12} to operate. But for a fault at location "m", the fault current is supplied by the grid as well as the DG-1. If the DG-1 supplies significant amount of fault current to the fault, it results in tripping of R_{13}, which is an unnecessary tripping of the relay R_{13}. This is also called as sympathetic tripping.

10.5.5   Loss of Mains (Islanding Condition)

Loss of mains indicates that the microgrid is operating in islanded mode of operation. In the test system shown in Figure 10.4, if the C.B. 2 opens, it indicates the DG-1 and DG-2 with some loads are isolated from the main

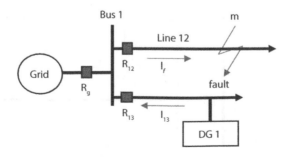

Figure 10.3 False/unnecessary tripping of relays.

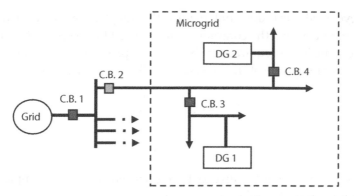

Figure 10.4 Loss of mains.

grid. It is called as an islanded condition, where the loss of mains occurred. If the microgrid is islanded, then resynchronization is a major concern since the utility grid's voltage magnitude, frequency and phase angle are different from the islanded grid. Loss of control over the voltage and frequency in the islanded system is also another concern.

In order to overcome the challenges, several protection schemes/techniques are developed by several researches around the globe, few of that are discussed in the following section.

10.6   Microgrid Protection

The various schemes developed for the microgrid protection are explained in this section.

10.6.1   Overcurrent Protection

Overcurrent protection is the widely used protection scheme in the protection system. In the conventional distribution system, only the non-directional overcurrent relays are utilized. Because, the power flow is always unidirectional and there is no reverse power flow. However, the power flow is bi-directional in microgrid, it is essential to have a directional overcurrent relay (DOCR). It is desired to have two different settings for the forward and reverse power flow for the DOCR. The dual settings for DOCRs are implemented in [12, 13], dual settings offer two different settings for forward and reverse power flow.

A dual setting DOCR relay with the assistance of communication has been developed in [12] for both the GCM and islanded mode of operation.

The developed work does not require the adaptive features for different modes of operation. The communication is used for the backup protection and the same is not required for the primary protection of the DOCRs.

The operating time of the inverse time DOCR can be expressed as

$$t_{ij} = TDS_i \frac{A}{\left(\dfrac{I_{scij}}{I_{pi}}\right)^B - 1}$$

The operating time of the inverse time dual setting DOCRs implemented in [12] is expressed as follows:

$$t^p_{fwij} = TDS_{fwi} \frac{A}{\left(\dfrac{I_{scij}}{I_{p\,fwi}}\right)^B - 1}$$

$$t^b_{rvij} = TDS_{rvi} \frac{A}{\left(\dfrac{I_{scij}}{I_{prvi}}\right)^B - 1}$$

where

    A and B – constants
    TDS – time dial setting
    fw and rv – forward direction and reverse direction
    I_{sc} and I_p – fault current and pick-up current
    i – relay identifier
    j – fault location

The main demerit of OCR is that the difference in fault current between the grid-connected mode and islanded mode is extremely significant, when CIDGs are utilized in the distribution network. This makes the coordination of OCRs very complex or impossible. In addition to that, the intermittent nature of renewable-based ICDGs makes the fault current level variable which further complicates the coordination of OCR further.

10.6.2   Distance Protection

A suitable alternative to OCR is a distance relay. The distance relay is not severely affected by the very less magnitude of fault current [11].

The line impedance is proportional to the line length; hence, it gives the distance measurement directly. So, it is apt to use impedance as a parameter for a relay to detect fault. Such a type is called as distance relay. In general distance relay is designed to detect fault from the relay location to a pre-defined location. Distance relay offers very high selectivity and generally used in transmission lines.

The impedance seen by the distance protection can be obtained as follows:

$$Z_l = \frac{V_k}{I_k}$$

During the normal operating condition, the Z_l is higher which is designed for the rated voltage and current. At the time of fault condition, V_k (voltage at the relay location) decreases and the I_k (current at the relay location) increases, results in reduction in Z_l. The reduction in impedance value indicates the fault condition.

In other words, in normal operating condition, impedance seen by any distance relay includes the load impedance which is higher in value. On the contrary, during the fault condition, the impedance seen by the distance relay is just the line impedance, i.e., the distance between the relay and fault location which is very lesser in value. Hence, on comparing the calculated value with the pre-defined value the fault within the protection zone can be detected easily.

The distance protection generally has three zones of protection, where the first zone covers 80% to 90% of the line to be protected. The second zone covers entire portion of the line to be protected and also 50% of the adjacent line. The third zone covers the 100% of the line to be protected and also 120% of the adjacent line [14]. The basic block diagram of distance protection is given in the Figure 10.5.

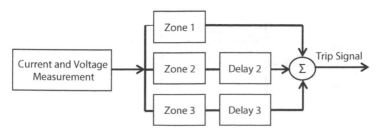

Figure 10.5 Basic block diagram of distance protection.

10.6.2.1   Effect of Distributed Generator Inclusion in the Distribution System on Distance Relay

In the system shown in Figure 10.6, if a fault occurred at location "m", then the impedance seen by the relay R_{12} is given by

$$Z_{measured\,1} = \frac{V_{12}}{I_{12}} = \frac{Z_{12}I_{12} + Z_m I_{23}}{I_{12}}$$

In the system shown in Figure 10.7, if a fault occurred at location "m", then the impedance seen by the relay R_{12} is given by [15]

$$Z_{measured\,2} = \frac{V_{12}}{I_{12}} = \frac{Z_{12}I_{12} + Z_m(I_{23} + I_{DG-1})}{I_{12}}$$

If the impedance seen by the R_{12} is compared for without and with DG conditions, then $Z_{measured\,2} > Z_{measured\,1}$. The integration of DGs in the distribution system increases the impedance seen by any particular distance relay in the system. Higher the capacity and current contribution of the DG, higher the impedance seen by the distance relays.

Dewadasa *et al.* [16] developed a novel inverse time characteristic (ITC) relay based on the measurement of admittance of the line to be protected.

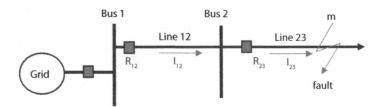

Figure 10.6 Test system without distributed generator.

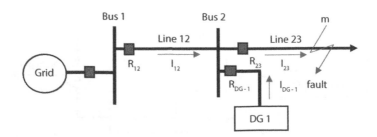

Figure 10.7 Test system with distributed generator.

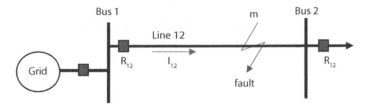

Figure 10.8 Test system for ITC.

Consider, the line L_{12} is to be protected in the test system shown in Figure 10.8. A fault occurs at a distance "m" from the relay location.

The total admittance of the line to be protected is Y_t and Y_m is the admittance from the relay location and the location of the fault. Then, the normalized admittance Y_r can be calculated as follows:

$$Y_r = abs\left(\frac{Y_m}{Y_t}\right)$$

The Y_r is utilized to obtain the ITC for the relay. In general, the ITC of the relay is given as

$$t_p = \frac{A}{Y_t^p - 1} + k$$

where

t_p – tripping time and
A, ρ, and k – constants.

10.6.3   Differential Protection

The differential protection senses the input and output current of the element to be protected. The difference in the current between the input and output element makes the OCR to operate. The basic block diagram of the differential relay is depicted in Figure 10.9. The differential protection comes under the category of unit protection [14].

Other protection schemes are affected by the following issues:

(i)   Bi-directional power flow
(ii)   fault current level variation due to the sporadic nature of DGs
(iii) Change in system configuration (radial to mesh)

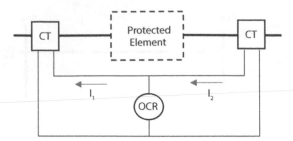

Figure 10.9 Basic block diagram of differential protection.

(iv) Fault current reduction in the islanded mode of operation
(v) Insignificant fault current variation at different relay due to short line lengths

These above-mentioned issues can be solved using differential protection. In order to protect a feeder using differential protection scheme, each end of the feeder is provided with a relay unit. Each relay unit comprises of phase [17–19].

10.6.3.1   Drawbacks in Differential Protection

Even though, differential protection is having very high selectivity for internal faults, does not affect due to the system or components rating. It needs a steadfast communication network in between the terminals of the protected component for the data transfer. Hence, it demands a backup protection scheme during a possible communication network failure, which results in increased system size and cost. This makes the differential protection schemes restricted in microgrids.

Although differential protection offers various advantages for protecting the entire microgrid, using differential protection is a costly affair.

In order to reduce the overall cost owing to the implementation of differential protection, the following measures can be implemented:

(i) sensors can be adopted instead of instrument transformers
(ii) differential protection can be used with other types of protection schemes.

10.6.4   Hybrid Tripping Relay Characteristic

A novel hybrid tripping relay characteristic is implemented in [20]. The tripping time is decided by both the current and voltage measurements; thus, it is called as hybrid tripping characteristics. The time of operation

(t_{op}) of the hybrid tripping relay characteristics is expressed using the following equation.

$$t_{op} = TMS\left(\frac{A}{\left(\frac{I_f}{I_p}\right)^B - 1} + C\right)\left(\frac{\alpha}{1-\left(\frac{V_f}{V_s}\right)^B} + \gamma\right)$$

where

A, B, C, α, β, and γ – constants
I_f and I_p – fault current and pick-up current
V_f and V_s – voltage at fault location during the fault and setting voltage

The prominent advantages of this method are as follows: it eliminates the need of communication, and it avoids the adaptive relay setting by fixing the same relay settings for both the islanded mode and the GCM. The reduction/low fault current situation is differentiated from the overloading condition by the voltage measurement. Moreover, it assists in reduction of operating time in the islanded mode. The proposed work eliminates the drawbacks in the works whose tripping time is solely depending on the fault current, where the tripping time increases as the fault current reduces. It is avoided by using voltage measurement also in the tripping time. The implemented method works efficiently in both low and high fault current scenarios and also able to discriminate the overloading condition from the low fault current.

10.6.5 Voltage-Based Methods

A voltage-based protection scheme has been developed in [21]; it is based on the direct and quadrature axis voltage in rotating reference frame. In the developed method, the measured three-phase voltage is converted to direct and quadrature axis in stationary reference frame, and it is further converted to d and q axes voltage in synchronously rotating reference frame. The same is done using the following equations.

$$\begin{bmatrix} v_{ds} \\ v_{qs} \\ v_0 \end{bmatrix} = \frac{2}{3}\begin{bmatrix} 1 & -\frac{1}{2} & -\frac{1}{2} \\ 0 & -\frac{\sqrt{3}}{2} & \frac{\sqrt{3}}{2} \\ \frac{1}{2} & \frac{1}{2} & \frac{1}{2} \end{bmatrix}\begin{bmatrix} V_a \\ V_b \\ V_c \end{bmatrix}$$

Table 10.1 Various signal features and the types of disturbance.

Sl. no.	Type	V_{dist} - Signal feature
1.	Three-phase disturbance	Pure DC voltage
2.	Phase-to-phase disturbance	DC voltage with AC ripples
3.	Single-phase disturbance	Oscillates between zero and maximum value

$$\begin{bmatrix} V_{dr} \\ V_{qr} \end{bmatrix} = \begin{bmatrix} \cos(\omega t) & -\sin(\omega t) \\ \sin(\omega t) & \cos(\omega t) \end{bmatrix} \begin{bmatrix} V_{ds} \\ V_{qs} \end{bmatrix}$$

The V_{dr} and V_{qr} are DC values. The change in voltage (V_{dist}) from the reference is obtained and the same is used to identify the disturbance and with the consecutive analysis offers the particulars about the type and location of fault.

$$V_{dist} = V_{q\,ref} - V_q$$

The V_{dist} is filtered in order to remove the undesired frequency components and is further processed in dual hysteresis controller. The minimum and maximum limits are helpful in obtaining the sensitivity of the fault identification.

The features of the signals are utilized to distinguish the different types of faults. The signal features and the types of disturbance are mentioned in the Table 10.1.

However, the developed method uses communication link between the relays to distinguish between the in-zone and out of zone faults.

10.6.6   Adaptive Protection Methods

Adaptive protection methods are the most emerging concepts in protection [22]. The relay settings may be changed according to the GCM or islanded mode or network topology. In general, the settings would be already calculated in offline and would be available as table. Any particular operating condition will be having its specific settings, so depending on the mode of operation and the network topology, that particular setting will be chosen. This method seems to be effective; however, communication is the backbone of such methods. The difficulty in protection when there is a communication failure is still a major concern. The communication security is yet to be explored.

10.7   Literature Survey

Few more literature papers are discussed in the Table 10.2.

Table 10.2 Pros of pros and cons of literature papers.

Ref.	Protection scheme	Remarks/pros	Cons
[23] [24]	Voltage-based method	Capable of identifying the internal and external faults to any protection zone.	Not validated for HIFs, 1 φ fault and 3 φ faults.
[16] [25]	Admittance-based method	Each downstream relay is protected by the upstream relay.	Suitability of the method in large system is yet to be verified.
[26]	Adaptive protection scheme	Modifies the settings according to the system configuration. Speed of operation is made possible by the communication. It also improves the selectivity.	HIFs are not considered, inclusion of new loads and DGs are not considered. The suitability of method due to the communication failure is a major concern for these types of methods.
[27]	Harmonic-based method	Penetration of DGs are allowed as long as harmonic contents are within the limits. This method is more suitable for CIDGs.	The suitability of the method for the inertial-based DGs are yet to be verified
[28]	Protection with optimal sizing and siting of Synchronous generation	A method is proposed for optimal siting and sizing of DGs, without modifying the existing relay settings. FCLs have also been incorporated when and where possible to make the proposed method effective.	Only synchronous-based DGs are considered. Inclusion of renewable energy-based DGs are not yet verified.

(Continued)

Table 10.2 Pros of pros and cons of literature papers. (*Continued*)

Ref.	Protection scheme	Remarks/pros	Cons
[29]	DOCR	The protection coordination for the GCM and the islanded modes are developed without the need of communication.	Nonetheless, DOCR based on programmable microprocessor is required.
[30]	Thevenin equivalent circuit parameter	According to the system configuration, the fault currents is calculated and Thevenin equivalent circuit parameter is estimated are in online, which are used for relay settings. It is applicable to both inertial and non-inertial–based DGs.	The complexity of this method increases as the network size increases. An additional offline computation is required to fix the relay setting.
[31]	Local measurement–based Non-unit protection	The impedance of the DC microgrid is calculated using the local measurement. This method is simpler and works faster.	The method is yet to tested for different topologies. At times it may be difficult to distinguish between the transient from the fault condition. The work is tested for DC microgrid only and yet to be tested for AC microgrid.
[32]	Symmetrical component and residual current–based scheme	Without communication this work provides protection to LL and LG faults.	Single pole tripping, HIF and three-phase faults are yet to be verified.
[33]	Travelling waves-based method	For fault identification and location, bus voltages and current travelling waves respectively are utilized	It is posted several advantages over the existing methods; however, the validation is not done.

(*Continued*)

Table 10.2 Pros of pros and cons of literature papers. (*Continued*)

Ref.	Protection scheme	Remarks/pros	Cons
[34] [35]	Pattern recognition–based scheme	Able to detect all types of faults including the high impedance faults	Threshold selection may be difficult to decide which plays a key role in the work.
[36]	Digital relays	Digital relays are utilized along with the proper communication network.	Higher capital investment, failure of the communication makes the protection system challenging.

10.8   Comparison of Various Existing Protection Schemes for Microgrids

The comparison of various existing protection schemes for microgrids are given in Table 10.3 [37–40].

10.9   Loss of Mains (Islanding)

As shown in Figure 10.10, a microgrid can be islanded in two ways, namely, planned (intentional) islanding and unplanned (unintentional) islanding. A typical microgrid is shown in Figure 10.11 [37–42].

Table 10.3 Comparison of various existing protection schemes for microgrids.

Protection scheme	Pros	Cons
Overcurrent	Simplest to implement and cheaper	1) Very difficult to distinguish between the short circuit fault current and the current at the transient condition. 2) Scheme becomes highly ineffective when CIDGs are used in the islanded system. 3) Highly affected by the network topology. 4) Source impedance need to be considered.

(*Continued*)

Table 10.3 Comparison of various existing protection schemes for microgrids. (*Continued*)

Protection scheme	Pros	Cons
DOCR	Suitable for protection in bidirectional power flow condition	1) Coordination of relays is very tedious for this type of relay. 2) Similar to OCR, it is very difficult to distinguish between the short circuit fault current and the current at the transient condition. 3) Source impedance need to be considered.
Distance	Source impedance variation is need not to be considered	The performance of the scheme is highly influenced by the bi-directional power flow, power swing, high impedance fault (HIF), high resistive fault, etc.
Differential	Simple, accurate, not influenced by the external fault, power swing and HIF, high selectivity and sensitivity	1) Communication transition delay and incorrect current measurements leads to ineffective of the relay performance. 2) This scheme is very costlier.
Voltage-based methods	Easy to classify the type and location of the fault	1) In a short line, a slight change in the voltage value makes a significant effect in the protection. 2) Dynamic network topology plays a vital role in this scheme.

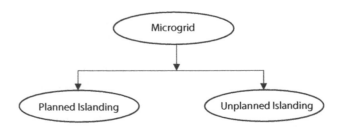

Figure 10.10 Ways of islanding formation.

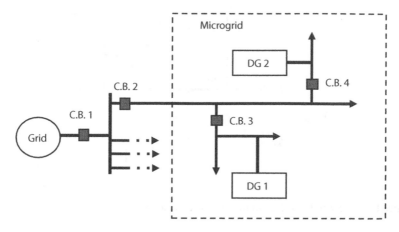

Figure 10.11 Typical microgrid.

Planned islanding is done for maintenance purpose, etc., whereas unplanned islanding is an accidental islanding. It may be occurred due to some electrical faults, mal-operation of relays or may be due to some unknown reasons. The reasons for the unplanned islanding are not required; however, it is necessary to detect the unplanned islanding according to several international standards.

10.10   Necessity to Detect the Unplanned Islanding

The necessity of detecting the unplanned islanding formation is given in the following sections [41–43].

10.10.1 Health Hazards to Maintenance Personnel

Consider, for some reason, the breaker C.B. 2 is opened. A maintenance crew is dispatched to perform the maintenance. In this scenario, the maintenance crew is unaware that the distribution networks are still energized and leads to permanent damage to working personnel or death. For a fault at the location "m" as shown in Figure 10.12, the C.B. 2 should open before C.B. 3 and C.B. 4. In case C.B. 2 is not opened before C.B. 3 and C.B. 4, in this condition also, the maintenance crew is unaware that the lines are still energized. Hence, this is a serious concern in unplanned island formation.

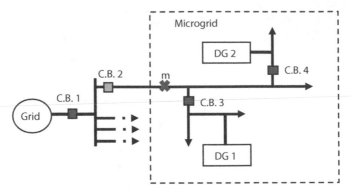

Figure 10.12 Health hazardous to maintenance personnel.

10.10.2   Unsynchronized Reclosing

Consider in the system shown in Figure 10.13, if the breaker C.B. 2 is opened for some temporary fault, then the microgrid and the rest of the system are separated. If C.B.1 is a recloser, it will try to reclose after a certain time period. This may lead to unsynchronized reclosing of the circuit breaker. The unsynchronized reclosing causes enormous amount of current which eventually damages the equipment in the distribution network and also in the utility grid.

10.10.3   Ineffective Grounding

Consider in the system shown in Figure 10.13, if the C.B. 2 is opened the microgrid is isolated from the main grid. At this operating condition, the effective grounding is lost with the utility grid. For a fault associated with ground results in more than 30% of voltage rise in the distribution network. It may damage the insulation in the distribution network equipment.

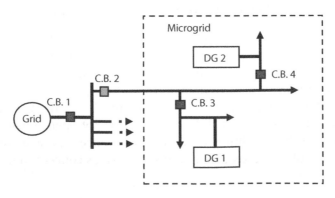

Figure 10.13 Unsynchronized reclosing.

Table 10.4 Various standards for ceasing operation of DGs on island detection.

Standard	Quality factor, (Q$_f$)	Islanding detection time, (t in s)
IEC 62116	1	t < 2 s
IEEE 1547	1	t < 2 s
Korean Standard	1	t < 0.5 s
UL 1741	1.8	t < 2 s
VDE 0126-1-1	2	t < 0.2 s
IEEE 929-2000	2.5	t < 2 s

10.10.4   Inept Protection

As discussed in the earlier sections, the fault currents differ in the microgrid during the islanded condition. It may result in the miscoordination of the over-current relays implemented in the system or non-detection of fault conditions.

10.10.5   Loss of Voltage and Frequency Control

Consider in the system shown in Figure 10.13, if the C.B. 2 is opened the microgrid is isolated from the main grid. During the GCM of operation the voltage and frequency are decided by the utility grid, upon island formation there is no control over the voltage or frequency on the post islanding condition. Because the DGs are unaware, whether its operating in GCM or islanded mode. It is too difficult for the DG to know whether it is operating in GCM or islanded mode, if the load is comparatively equal to the DG power generation.

Due to the aforementioned reasons, it is essential to identify the unplanned island formation as suggested by the various international standards within the stipulated time as given in Table 10.4.

10.11   Unplanned Islanding Identification Methods

The identification of the unplanned island formation is called as unplanned islanding identification methods (UIIMs). The UIIMs are broadly divided into two types, *viz.*, communication-based methods (remote methods) and non-communication–based methods (local techniques). The different types of UIIM are depicted in Figure 10.14 [41–43].

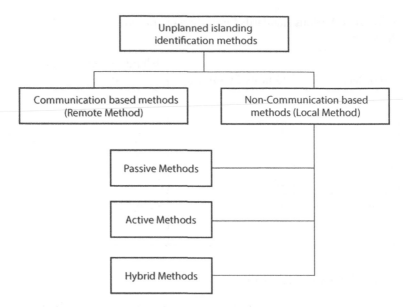

Figure 10.14 Different types of UIIM.

10.11.1   Communication-Based Methods (Remote Method)

This method uses communication networks (wired or wireless) for transferring the status of the substation circuit breaker to the DG site [42]. Since it uses communication, the accuracy of the islanding identification is very high. However, the capital investment and the security of the signal remains a major concern. It is not preferred for low-rated microgrids owing to its higher capital investment.

10.11.2   Non-Communication–Based Methods (Local Method)

This method uses the measured parameters available at the DG terminal to identify and differentiate the islanding and non-islanding conditions. This method is again divided into passive, active, and hybrid methods.

10.11.2.1   Passive Method

It is the simplest method of identifying the island formation. It measures the parameters (voltage, frequency, power angle, etc.) at the DG terminal and compares the measured value with the pre-defined threshold. When the measured value exceeds the threshold, it is concluded as island formation. Even though it is the easiest method and does not degrade the power quality [42, 43].

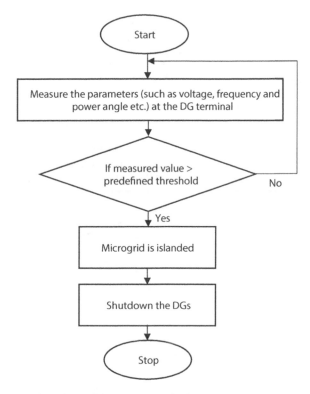

Figure 10.15 Basic flowchart of the passive method.

When difference between the DG generation power and the load power is less, this method fails to detect the island formation and the operating zone where the methods mail to identify the island formation is called as Non-Detection Zone (NDZ). The basic flowchart of the passive method is shown in Figure 10.15.

10.11.2.2   Active Method

This method is developed to nullify or reduce the drawbacks in the passive method. In this method, a disturbance is injected deliberately into the system or the inverter control is appended with positive or negative feedback. Due to the disturbance or feedback, the voltage magnitude or frequency or some other parameters varies significantly from the nominal value on post islanding condition. Thus, the non-detection is either completely eliminated by the active method or reduced significantly than the passive method. However, this method degrades the power quality in the GCM of operation which is not desirable. Some methods may destabilize the system on post islanding conditions which

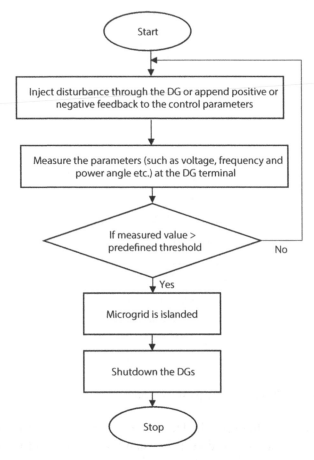

Figure 10.16 Basic flowchart of the active method.

is also not desirable. The basic flowchart of the active method is shown in Figure 10.16 [42, 43].

10.11.2.3 Hybrid Method

This method is developed to reduce the drawbacks in both the passive method and active method. In this method, passive method is used to suspect the island formation and active method is used to identify the island formation. Both the methods are implemented one after another. Thereby, the non-detection zone in the passive method is reduced by using the active method and since active method is implemented only after the suspect by the passive method. It completely eliminates the power quality degradation during the GCM of operation. The basic flowchart of the hybrid method is shown in Figure 10.17 [44, 45].

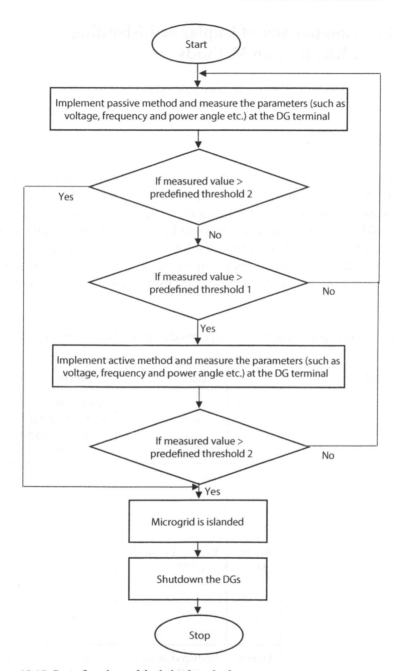

Figure 10.17 Basic flowchart of the hybrid method.

10.12   Comparison of Unplanned Islanding Identification Methods

The comparison of UIIMs is given in Table 10.5.

10.13   Discussion

Even though several works have been developed for microgrid protection and UIIM, there is hardly any technique which can work effectively for all operating conditions and all topologies, there are several issues need to be considered to make an effective technique. Adaptive protection seems to be a better option or mostly suited for microgrid operation. However, communication dependency makes the method bit difficult due to the possibility of failure. The adaptive protection using a local measurement may

Table 10.5 Comparison of various unplanned islanding identification methods [37–45].

Method		Merits	Demerits
Communication-based methods (remote)		Identification accuracy is higher	High cost, Security and delay are also major concern
Non-communication–based methods (local)	Passive	Cheaper and does not degrade the power quality	Higher NDZ
	Active [46]	Reduced or zero NDZ	Degrades the power quality during GCM of operation, Costlier than passive method
	Hybrid	Reduced or zero NDZ	Islanding detection time may be higher than passive and active method

be a better solution in terms of protection. UIIM is essential according to various standards. Hybrid methods for UIIM may be suitable solution for the reduced or zero NDZ with low or zero power quality degradation during GCM.

10.14   Conclusion

Despite numerous advantages of the microgrid, the main challenge is the microgrid protection. The microgrid protection is different from the conventional protection owing to the following reasons, *viz.*, a) current is no more unidirectional, b) topology may vary according to the necessity, c) change in fault current due to the varying nature of energy resources input in the distribution system, d) insignificant fault current contribution in the islanded system, and e) insignificant fault current difference due to the shorter lengths. Due to the aforementioned reasons, the conventional protection system is not effective. Hence, it is essential to develop a novel protection scheme which works effectively for all the operating conditions and topologies.

References

1. Lasseter, B., Microgrids [distributed power generation], in: *2001 IEEE Power Engineering Society Winter Meeting. Conference Proceedings (Cat. No.01CH37194)*, vol. 11, pp. 146–149, 2001.
2. Lasseter, R.H., Microgrids, in: *2002 IEEE Power Engineering Society Winter Meeting. Conference Proceedings (Cat. No.02CH37309)*, vol. 1, pp. 305–308, 2002.
3. Venkataramanan, G. and Illindala, M., Microgrids and sensitive loads, in: *2002 IEEE Power Engineering Society Winter Meeting. Conference Proceedings (Cat. No.02CH37309)*, vol. 1, pp. 315–322, 2002.
4. Pesin, M., U.s. department of energy electricity grid research and development. Presentation at the *American Council of Engineering Companies, Environment and Energy Committee Winter Meeting*, 9 Feb 2017.
5. Cigré, Working group C6.22. Microgrids 1 Engineering, Economics, & Experience, in: *CIGRE session Technical Brochure*, Oct 2015.
6. Patnaik, B., Mishra, M., Bansal, R.C., Jena, R.K., Ac microgrid protection – a review: Current and future prospective. *Appl. Energy*, 271, 115210, 2020.
7. Gopalan, S.A., Sreeram, V., Iu, H.H., A review of coordination strategies and protection schemes for microgrids. *Renewable Sustainable Energy Rev.*, 32, 222–228, 2014.

8. Barra, P., Coury, D., Fernandes, R., A survey on adaptive protection of microgrids and distribution systems with distributed generators. *Renewable Sustainable Energy Rev.*, 118, 109524, 2020.

9. Parhizi, S., Lotfi, H., Khodaei, A., Bahramirad, S., State of the art in research on microgrids: A review. *IEEE Access*, 3, 890–925, 2015.

10. Mirsaeidi, S., Mat Said, D., Wazir Mustafa, M., Hafiz Habibuddin, M., Ghaffari, K., An analytical literature review of the available techniques for the protection of micro-grids. *Int. J. Electr. Power Energy Syst.*, 58, 300–306, 2014.

11. Hooshyar, A. and Iravani, R., Microgrid protection. *Proc. IEEE*, 105, 7, 1332–1353, 2017.

12. Zeineldin, H.H., Sharaf, H.M., Ibrahim, D.K., El-Zahab, E.E.A., Optimal protection coordination for meshed distribution systems with dg using dual setting directional over-current relays. *IEEE Trans. Smart Grid*, 6, 1, 115–123, 2015.

13. Sharaf, H.M., Zeineldin, H.H., El-Saadany, E., Protection coordination for microgrids with grid-connected and islanded capabilities using communication assisted dual setting directional overcurrent relays. *IEEE Trans. Smart Grid*, 9, 1, 143–151, 2018.

14. Paithankar, Y. and Bhide, S., *Fundamentals of power system protection*, PHI Learning Private Limited, India, 2010.

15. Lin, H., Liu, C., Guerrero, J.M., Vásquez, J.C., Distance protection for microgrids in distribution system, in: *IECON 2015 - 41st Annual Conference of the IEEE Industrial Electronics Society*, pp. 731–736, 2015.

16. Dewadasa, M., Majumder, R., Ghosh, A., Ledwich, G., Control and protection of a microgrid with converter interfaced micro sources, in: *2009 International Conference on Power Systems*, pp. 1–6, 2009.

17. Dewadasa, M., Ghosh, A., Ledwich, G., Protection of microgrids using differential relays, in: *AUPEC 2011*, pp. 1–6, 2011.

18. Sortomme, E., Ren, J., Venkata, S.S., A differential zone protection scheme for microgrids, in: *2013 IEEE Power Energy Society General Meeting*, pp. 1–5, 2013.

19. Wheeler, K.A., Faried, S.O., Elsamahy, M., A microgrid protection scheme using differential and adaptive overcurrent relays, in: *2017 IEEE Electrical Power and Energy Conference (EPEC)*, pp. 1–6, 2017.

20. Chakraborty, S. and Das, S., Communication-less protection scheme for ac microgrids using hybrid tripping characteristic. *Electr. Power Syst. Res.*, 187, 106453, 2020.

21. AL-Nasseri, M.A. and Redfern, H., A New Voltage based Relay Scheme to Protect Micro-Grids dominated by Embedded Generation using Solid State Converters. *19th International Conference on Electricity Distribution*, no. 0723, pp. 1–4, 2007.

22. Habib, H.F., Lashway, C.R., Mohammed, O.A., On the adaptive protection of microgrids: A review on how to mitigate cyber attacks and communication failures, in: *2017 IEEE Industry Applications Society Annual Meeting*, pp. 1–8, 2017.

23. Al-Nasseri, H., Redfern, M.A., O'Gorman, R., Protecting micro-grid systems containing solid-state converter generation, in: *2005 International Conference on Future Power Systems*, p. 5, 2005.

24. Al-Nasseri, H., Redfern, M.A., Li, F., A voltage based protection for micro-grids containing power electronic converters, in: *2006 IEEE Power Engineering Society General Meeting*, p. 7, 2006.

25. Majumder, R., Dewadasa, M., Ghosh, A., Ledwich, G., Zare, F., Control and protection of a microgrid connected to utility through back-to-back converters. *Electr. Power Syst. Res.*, 81, 7, 1424–1435, 2011.

26. Laaksonen, H.J., Protection principles for future microgrids. *IEEE Trans. Power Electron.*, 25, 12, 2910–2918, 2010.

27. Pandi, V.R., Zeineldin, H.H., Xiao, W., Determining optimal location and size of distributed generation resources considering harmonic and protection coordination limits. *IEEE Trans. Power Syst.*, 28, 2, 1245–1254, 2013.

28. Srividhya, S. and Murali, V., Optimal siting and sizing of distributed generators to retain the existing protection setup along with their technical and economic aspects. *IET Gener. Transm. Distrib.*, 13, 15, 3240–3251, 2019.

29. Zamani, M.A., Sidhu, T.S., Yazdani, A., A protection strategy and microprocessor-based relay for low-voltage microgrids. *IEEE Trans. Power Deliv.*, 26, 3, 1873–1883, 2011.

30. Shen, S., Lin, D., Wang, H., Hu, P., Jiang, K., Lin, D., He, B., An adaptive protection scheme for distribution systems with dgs based on optimized thevenin equivalent parameters estimation. *IEEE Trans. Power Delivery*, 32, 1, 411–419, 2017.

31. Meghwani, A., Srivastava, S.C., Chakrabarti, S., A non-unit protection scheme for dc microgrid based on local measurements. *IEEE Trans. Power Delivery*, 32, 1, 172–181, 2017.

32. Nikkhajoei, H. and Lasseter, R.H., Microgrid protection, in: *2007 IEEE Power Engineering Society General Meeting*, pp. 1–6, 2007.

33. Shi, S., Jiang, B., Dong, X., Bo, Z., Protection of microgrid, in: *10th IET International Conference on Developments in Power System Protection (DPSP 2010)*, Managing the Change, pp. 1–4, 2010.

34. Samantaray, S.R., Joos, G., Kamwa, I., Differential energy based microgrid protection against fault conditions, in: *2012 IEEE PES Innovative Smart Grid Technologies (ISGT)*, pp. 1–7, 2012.

35. Kar, S. and Samantaray, S.R., Time-frequency transform-based differential scheme for microgrid protection. *IET Gener. Transm. Distrib.*, 8, 2, 310–320, 2014.

36. Sortomme, E., Venkata, M., Mitra, J., Microgrid protection using communication-assisted digital relays, in: *IEEE PES General Meeting*, pp. 1–1, 2010.

37. Ku Ahmad, K.N.E., Selvaraj, J., Rahim, N.A., A review of the islanding detection methods in grid-connected pv inverters. *Renewable Sustainable Energy Rev.*, 21, 756–766, 2013.

38. Suman, M. and Kirthiga, M.V., Unintentional islanding detection, in: *Distributed Energy Resources in Microgrids*, R.K. Chauhan and K. Chauhan (Eds.), pp. 419–440, Academic Press, London, United Kingdom, 2019, ch. 17.

39. Murugesan, S. and Murali, V., Disturbance injection based decentralized identification of accidental islanding. *IEEE Trans. Ind. Electron.*, 67, 5, 3767–3775, 2020.

40. Khamis, A., Shareef, H., Bizkevelci, E., Khatib, T., A review of islanding detection techniques for renewable distributed generation systems. *Renewable Sustainable Energy Rev.*, 28, 483–493, 2013.

41. Murugesan, S. and Murali, V., Band pass filter and afvmean-based unintentional islanding detection. *IET Gener. Transm. Distrib.*, 13, 9, 1489–1498, 2019.

42. Li, C., Cao, C., Cao, Y., Kuang, Y., Zeng, L., Fang, B., A review of islanding detection methods for microgrid. *Renewable Sustainable Energy Rev.*, 35, 211–220, 2014.

43. Murugesan, S. and Murali, V., Q-axis current perturbation based active islanding detection for converter interfaced distributed generators. *Turk. J. Elec. Eng. Comp. Sci.*, 26, 5, 2633–2647, 2018.

44. Laaksonen, H., Advanced islanding detection functionality for future electricity distribution networks. *IEEE Trans. Power Delivery*, 28, 4, 2056–2064, 2013.

45. Kermany, S.D., Joorabian, M., Deilami, S., Masoum, M.A.S., Hybrid islanding detection in microgrid with multiple connection points to smart grids using fuzzy-neural network. *IEEE Trans. Power Syst.*, 32, 4, 2640–2651, 2017.

46. Murugesan, S. and Murali, V., Active unintentional islanding detection method for multiple pmsg based dgs. *IEEE Trans. Ind. Appl.*, 56, 5, 4700–4708, 2020.

11

Microgrid Optimization and Integration of Renewable Energy Resources: Innovation, Challenges and Prospects

Blesslin Sheeba T.[1], G. Jims John Wessley[2], Kanagaraj V.[3], Kamatchi S.[4], A. Radhika[5] and Janeera D.A.[5*]

[1]*Department of ECE, R.M.K. Engineering College, Kavaraipettai, India*
[2]*Department of Aerospace Engineering, Karunya Institute of Technology and Sciences, Coimbatore, India*
[3]*Department of Electrical and Electronics Engineering, University of Technology and Applied Sciences-Al Mussanah, Muladdah, Oman*
[4]*Amrita School of Engineering, Bangalore, India*
[5]*Sri Krishna College of Engineering and Technology, Coimbatore, India*

Abstract

Renewable energy is creating a growing interest on a global level. Integration of renewable energy resources into the power grid structure is challenging due to their intermittent nature. Advanced solutions for connection of renewable energy sources (RES) are investigated with the increase in the number of RES. Microgrids offer promising solution for this purpose. Microgrids provide heat and power to a local area through a single controllable system that consists of micro sources and cluster of loads. The single power source is transformed to multiple and bidirectional power sources due to the radial distribution system incorporated with renewable energy technologies. The radial distribution networks and its associated power outages are reduced and the reliability of the system is increased with this model. Various other factors such as power system stability, energy efficiency, demand side management, carbon dioxide emission, computer tools, and economic assessment are compared and the highest potential is identified.

Corresponding author: janeera@skcet.ac.in

M. Kathiresh A. Mahaboob Subahani and G.R. Kanagachidambaresan (eds.) Integration of Renewable Energy Sources with Smart Grid, (239–262) © 2021 Scrivener Publishing LLC

Keywords: Renewable energy, microgrid, integration, radial distribution networks, energy optimization

11.1   Introduction

Diverse challenges like economic recession, energy supply security, and climate change, have emerged in the society in the recent times [1]. These issues are addressed by the energy sector. The economic progress is significantly influenced by the conventional energy sources like natural gas, coal, and oil. However, the rapid industrialization, high living standards, and global economic expansion due to the growing population require higher energy consumption every year [2]. This leads to the need for depending on renewable energy sources (RESs). With the global environmental issues, depletion of fossil fuels, high cost of fossil fuels, concerns over climatic changes, rapid growth in power demand, and power system restructuring, the non-polluting and inexhaustible characteristics of RERs have put them in the forefront of sustainable energy development. A power system has to satisfy the power demands of all its customers with good quality and high reliability in an efficient and economic manner. For this purpose, RERs are integrated with the power systems in several countries [3].

The RER power output is expected to rise around 10% in a decade according to the World Energy Council (WEC) survey. The economic activities of a nation depend on the energy segment for boosting the economy and enabling economic development. Thus, reliability of power system is of prime significance [4]. During energy planning, identifying the prospective of renewable energy and plays a major role. In order to analyze the performance of the model that is designed, reliability is a major performance indicator. Other notable performance indicators include power system stability, energy efficiency, demand side management, carbon dioxide emission, computer tools, and economic assessment [5]. Implementation of renewable energy and the effect of renewable energy on the energy system is identified by coherent technical analysis. The power system performance of utilities can be measured at conceptual as well as operating stages.

According to certain survey, it is evident that 80% of power supply unavailability at load points is due to the distribution system failure. The sophisticated characteristics and sensitivity of the hardware causes an economic blow to the consumers as a localized effect. Hence, in power sector, the distribution system reliability is crucial [6]. Assessment of the reliability of the distribution system has to be performed frequently by the distribution network operators along with the customer failure surveys.

Revenue loss, raw materials spoilage, sales loss, and production loss may occur due to the persistent interruptions of power leading to socioeconomic impact on the consumers. Despite the availability of reliable power generation and transmission systems, consumers are greatly affected by the poor supply and distribution systems in several regions [7]. This indicates that the power distribution system reliability has to be assessed regularly to improve the efficiency of consumption.

At load points of the consumers, power outage frequency and duration can be greatly reduced by improving the power distribution system reliability [8]. In deregulated power environment, power regulatory systems can be furnished with significant information through reliability assessment. Traditional power distribution systems were dependent on a single power source. The entire system is affected, causing severe power interruption due to short-circuit fault or power failure on the network. The lack of alternative path to provide continuous power supply to consumers lead to power outages at the load points. In order to restore the power supply of the system, fault clearance must be performed in radial distribution systems due to their serial configuration [9]. The voltage regulation, total system length, and maximum power demand are the factors that decide the radial distribution system's optimal operation.

The consumers associated with the lateral feeders are affected due to the fault in the main feeders of radial distribution system. This leads to very low system reliability. When the system loads vary, persistent voltage fluctuation may occur at the consumer end of the distribution system [10]. Multiple communication facilities and power sources can be used when microgrids are used, and thus, the power system reliability can be improved. The annual cost of the system (ACS), net present cost (NPC), and cost of energy (COE) can be reduced and uninterrupted power supply may be provided when the load points and renewable energy distributed generation units are close to each other in microgrid system configuration [11]. Unexpected power outages are reduced and microgrid system reliability is improved invariably due to this configuration.

The application of RERs even in distant regions has increased due to the high-power demand without increasing the generation capacity and high transmission and distribution (T&D) cost of the system [12]. In terrain regions, the rural dwellers load requirements are met by RERs as they are the most cost-efficient power solutions and extension of T&D lines to these regions are expensive. Several international organizations have reported that about 17% of global population live without electricity. Remote communities in rural areas of developing nations with issues in grid connection due to various constraints has approximately 22% of people living without electricity. The load requirements at the consumer load points are met by incorporating RERs in the power system of such locations. This enhances the reliability of power system, increases the reserve margin, and reduces the associated operating cost of the system. Microgrid systems consisting of battery storage system (BSS), photovoltaic (PV), wind turbine generator (WTG), and other renewable energy models can be used for promoting the system utilization and meeting the sustainability of global energy [13].

The available wind and solar energy resources are utilized, thereby reducing the dependence on fossil fuel for power generation applications. Reduction in emission, operation, maintenance, and fuel cost as well as improved power system reliability has enabled RERs to be incorporated into several power systems on a global level. In order to improve the global electricity generation, RERs are utilized due to their unique characteristics. Exploitation of RERs has increased with the global awareness, environmental impact, and public awareness of depleting fossil fuels along with the exponential increase in power demand. The technical and economic benefits of wind and solar energy has led to various research and development activities in the domain [14]. Reduction in the emission of greenhouse gases (GHG) and climate change mitigation is enabled by using alternative power generation sources like wind and solar energy by the distribution network operators.

Incorporation of different power sources in a power system helps in reducing the challenges faced due to the stochastic characteristics of wind and solar systems to a certain extent. The weakness of a RES can be overcome by the strength of another resource in the power utilities [15]. Without depending on the battery storage system, the effects of local RER's intermittent nature can be can be addressed. During peak hours or insufficient power supply, the batter storage system can be operated. Economy and reliability of power system is assessed based on the effects of BSS and RER by several researchers. However, the power system reliability assessment in terms of the economic aspect and stochastic characteristics of the significant components of the system has not been covered in detail. The total outage cost (TOC), expected energy not served (EENS), and expected interruption cost (ECOST) index of the system is to be analyzed for assessment of power system reliability. This is because the radial nature of the distribution system network causes around 80% of power interruptions [16].

11.2   Microgrids

A microgrid is a cluster of local distributed energy resource (DER) and loads in such a way that it can operate in island mode or within the grid.

It is connected either at a low or medium voltage level [17]. The connected microgrid resembles a single node that can generate power or consume the grid power. The loads are categorized into reliability classes and the energy storage devices are used for operation in isolation mode. A simple microgrid with various storage devices and generation units is as represented in Figure 11.1. Various reliability classes can be used for classification of loads into categories like shed-able, adjustable, and sensitive.

Efficient energy utilization and enhancement of reliability are the key aspects of microgrids. This enables implementation of islanded operation. RES is commonly used for energy supply. Storage units and Information and Communication Technology (ICT) is essential for this purpose. Uncertainties in the production of electricity by RES can be balanced by demand side management. During isolated operation, stability during load connection or disconnection is challenging as the installed generation capacity is small in size. Yield problems may be faced at the operation costs due to the efficient energy utilization of microgrids [18].

Microgrids have very small inertia. An imbalance occurs between the mechanical torque at the input and electromagnetic torque at the output when a rotating generator exists in the system resulting in the occurrence of a fast acceleration or deceleration of the rotor. However, synchronized power electronic inverters may be used for connection of all generation units in microgrids. The rotating masses and frequency are independent of each other in this scenario. On occurrence of system imbalance, the control

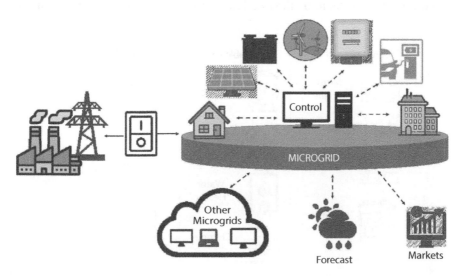

Figure 11.1 A simple microgrid platform.

system and the information should react quickly. Batteries can be charged if the power supply is large [19]. Power balance can be restored by switching off the shed-able loads when the system relies on the supply for power. When compared to the capacity of the installed generation, if the loads are large, in isolated operation, connection or disconnection of loads is crucial.

Reliability is a major issue in the microgrid approach. Microgrid offers high degree of reliability as it can operate in an isolated form. The microgrid can decouple and continue operation even if fault occurs in the connecting grid. In critical units like data backup centers and hospitals, this feature is significant. Inclusion of redundancy for critical units can improve the security and reliability of microgrids in islanded operation. In order to deal with the inconsistency of units that are dependent on weather, backup devices such as fuel generators or large storage capacity is required for the microgrids in islanded operation [20].

The loads and generation units can be of the same scale in islanded operation, thereby increasing their importance in the system. Demand side management also has a significant role in this model [21]. Based on the various types of reliability, loads are classified into shed-able, adjustable and sensitive type. Sensitive loads require continuous power supply and are of highest importance. When the power generation is insufficient at any time duration, shed-able loads can be detached. At a given power interval, control of adjustable loads can be done.

Figure 11.2 represents the load categorization on microgrids. The storage capacity can be reduced and weather dependent generation can be balanced by adjustable or flexible loads. In order to increase the system

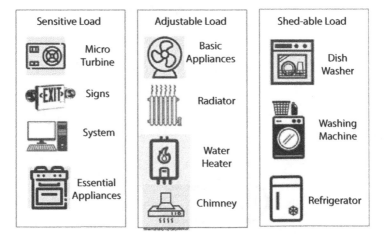

Figure 11.2 Load categorization on microgrids.

storage capacity, plugin hybrids are used. In islanded operation of micro-grids, energy efficiency plays a major role. Economic and efficient energy handling is required due to the limitations of power generation capacity. Energy efficiency can be improved by appropriate classification of loads [22]. The amount of energy purchased from the grid is reduced by the microgrid leading to lower consumption of total primary energy. Removal of transmission losses and use of combined heat and power (CHP) units contribute toward this energy reduction.

Lesser energy wastage in microgrids leads to reduction in the emission of CO_2. Lower emission, in turn, leads to reduction of total energy consumption. Certain approaches like gas turbines worsen the CO_2 balance in the atmosphere. In order to overcome this, integrated fossil fired generators are used along with virtual power plant approach. Rather than being generated in the microgrid, when electricity is purchased from the grid, the balance of CO_2 is better according to previous simulation results. Implementation of microgrids in smaller local areas offer better reliability. The probability of failure of islanding operation and fault occurrence within the microgrid is larger with increase in the microgrid line distances [23]. Redundancy can be increased in microgrids with several units. In order to increase the reliability of a microgrid, the number of units involved and the application area are to be optimized.

11.3   Renewable Energy Sources

In order to reduce the GHG emission and overall environmental damage, sustainable energy services can be provided by the RERs. They play a major role in the transformation of global energy consumption. Over the past few decades, the renewable energy power system costs have reduced substantially, making it more desirable for household and commercial purposes [24]. Several forecasts indicate further decline in the renewable energy generation cost. It also proves beneficial in terms of economy, ecology, and energy, making it widely sought after for on-site power generation applications. About 18% of the global energy demand is addressed by RES. In terms of off-grid rural supply, fuel transportation, space heating, hot water, and power generation, the conventional fuels are replaced by renewable energy. About 3.4% of global electricity generation is contributed by RESs. This scenario is depicted in Figure 11.3 based on the data provided by International Renewable Energy Agency (IRENA).

Geothermal energy, biogas, solid and liquid biofuels, renewable waste, solar power, solar PV, wind, and hydropower are some of the basic RES

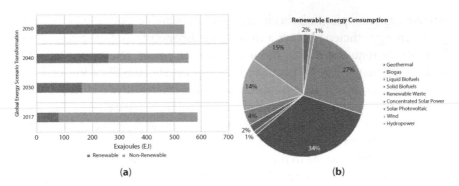

Figure 11.3 (a) Global energy scenario transformation. (b) Renewable energy consumption in USA based on renewable energy type.

available. Solar and wind energy are used in most of the systems that use renewable energy on a global scale [25]. In comparison with the conventional resources available in the power market, wind energy offers competitive power generation performance.

11.3.1   Renewable Energy Technologies (RETs)

Over the last few decades, several renewable power plants are installed. Small generators are utilized to supply electricity on site or attached to the distribution system. This model in microgrids are termed as the distributed power generation (DG). Reliability and quality of service are established for developed economies through efficient, secure, and clean energy system by integrating microgrids with DG technology [15]. Various promising load management, distributed storage (DS), and power generation systems are being developed for commercialization. These technologies enhance the microgrid infrastructure.

Some of the commonly used renewable energy technologies include microturbines, fuel cells, PV cells, solar thermal field, and wind turbines. A small gas turbine and generator mounted on a single shaft are the constituents of the microturbines. The size of these units ranges between 30 and 100 kW with 20%–30% efficiency. Overall efficiency of the system can be improved to up to 80% with competitive costs through CHP operation. Fuel cells enable high efficiency power conversion of 35%–55% without heat recovery [18]. Research and development are widely undertaken toward utilization of fuel cells for power generation. Phosphoric acid fuel cells are commonly used in microgrids. The overall fuel conversion efficiency is enhanced by CHP operation. Solid oxide, molten carbonate, proton exchange membrane, and several other fuel cell technologies are being developed.

Ever since the space program was introduced by U.S., PV devices are being used. At cell terminals, DC voltage is produced via sunlight [25]. The cell design and intensity of sunlight are the factors that determine the amount of current and voltage produced by the PV cells. Fixed or moving cell arrays are used for capturing the solar energy in PV cells. When available sunlight cannot meet the power requirements, stored power maybe used from the standalone systems. Zero emission, quietness, and high reliability of operation are the major advantages of PV systems.

Sterling dish is the major technology used in solar thermal field for small-scale power generation. The size of these units ranges between 10 and 25 kW. An array of sun-tracking mirrors is used in which the small receiver receives the concentrated light. Further, the gathered heat is transmitted to the sterling engine's hot end [26]. The shaft rotation is generated by pistons being pushed in closed cycle using the working fluid in sterling engine. The electric grid is connected to an induction generator that spins using the shaft rotation in the sterling dish. Wind-based power generation has been used commercially since several years. Wind turbines in large wind farms can produce 700 kW to 1.5 MW of power. The blades and rotor of an induction generator is used for driving these machines. Other power generation systems or storage system must provide the excess power requirement for turbines under stand-alone operation mode.

11.3.2   Distributed Storage Technologies

With load changes, the power generation sources of microgrids may not be able to make smooth transition. In such cases, utilization of purchased power during peak loads may be too pricey. Hence, storage plays a major role in the microgrid system. Short-lived fast or slow transient causes load changes in microgrids [3]. They may be caused by increase in ramping capability, equipment on/off operation, or due to starting of motors. When the utility power is lost, quick tracking and immediate back-up provision can be provided by the storage systems that can quickly transition into functioning in sub-cycle time frames.

The battery systems have extensive operational experience as they have been used for storage of electrical energy traditionally. Different sizes of lead-acid batteries provide back-up power to several commercial and house-hold applications. Flow batteries and other chemical batteries are also available in the market [27]. Battery lifetime and energy storage density has been enhanced with the developments in new batteries. Advanced power electronics, magnetic bearings, and composite rotors are incorporated in the flywheel system that can store high speed energy.

The generators or motors are connected to disks and rotors or rotating wheels can operate at a speed of up to hundred thousand rpm. The flywheel power storage capacity is directly proportional to the square of speed of rotation. Energy discharge can be performed by flywheels over long periods in a slower rate or quickly at high power in the range of kW.

Electrical current is passed by means of superconductors without any significant loss. The coil wire of superconductor serves as a path for circulating current which enables storage of electrical energy. The energy storing magnetic field is established by this circulating current [9]. The superconducting coil is charged and discharged by power electronic interfaces. Energy is stored as electrostatic charge in high capacity electrolytic devices termed as supercapacitors. A thin separator separates the two electrodes of the supercapacitors. With the increase in the electrodes surface area, the capacity of energy storage increases [11].

11.3.3   Combined Heat and Power

Sustainability of microgrids is ensured by incorporation of CHP systems for heat recovery. During the process of converting the primary fuel to electricity, waste heat is generated in the system. Significant benefits may be obtained by productive utilization of this heat in the microgrid. During power generation, the amount of unutilized primary energy that is released into the environment sums up to 50% to 75%. Transportation of heat in the form of hot water or steam over long distances, unlike electricity, is not economical [28]. For this reason, this heat is redirected by CHP for industrial purposes like sterilization, domestic hot water, local district heating, and on-site space heating.

The benefits of this system can be utilized completely by moving the heat production unit closer to the point of use. Relative to the heat loads, the generators can be placed optimally in the microgrids as electricity can be transported more readily when compared to heat. The heat requirements can be matched while providing improved flexibility as the individual heat production units are of smaller scale. The waste and non-waste heat production generators can be combined together in an economic manner for electricity generation and heat optimization in a microgrid [29].

11.4   Integration of RES in Microgrid

Energy storage system, fuel cell, wind energy, solar energy, and mini-hydro energy system are the major components of microgrid. Generation of electricity, storage of energy, and operation of load are performed by the integration of these components to the main (macro) grid. Microgrids can operate in stand-alone or grid-connected mode [13]. The isolated or standalone mode of operation is a major benefit of microgrid as it enables autonomous functioning of the system. Low-voltage interconnections are used for power generation and loads in microgrids. However, the operator should be cautious regarding the amount of power systems linked to the microgrid.

Highly reliable power supply can be provided by an isolated microgrid due to the energy sources like wind, solar, and fuel cells that are included for power generation. A flexible trade-off is allowed between the electric power and heat requirements due to the heat produced by microturbines and other generation sources which can be used for space or local process heating. A 30- to 50-km radius can be covered by small microgrids that can produce 5 to 10 MW of power for consumption [18]. Microgrids are independent of long-distance transmission line and hence free from huge transmission losses. DC power generation concept is adapted in the design of DC microgrids. Here, minimum distance must be maintained between the load and power generation source. The DC microgrid requirements can also be met by cost efficient wind or solar farms at a particular site. Minimal AC to DC or DC to AC conversion must take place.

The DC storage devices like fuel cells, capacitors, and batteries have to meet the local DC electricity requirements. A microgrid is a minor version of the huge power grid with the ability to operate on a self-sufficient energy network with power generation and storage devices. The traditional use of AC transmission lines is evolving with the technological advancements. The AC transmission line output and square of voltage is directly proportional to each other, whereas inversely proportional to the transmission line impedance [7]. As the distance between the load and the power source increases, the power loss increases. The voltage level can be increased in long distance AC transmission lines for attaining high transmission capability.

While implementing the microgrid system, based on the availability of input source and consumer requirements, the renewable energy resources for power supply system has to be selected. All the RERs must be integrated in the microgrid. Further, the energy storage, management and control system are to be employed. The environment and location of installation is crucial while selecting the RER for microgrid [23]. Efficient energy storage and management schemes are to be implemented to keep in pace with the technological advancements, overcome environmental concerns related to the use of fossil-based fuel, employ distributed energy sources, increase RES utilization, increase electric vehicle utilization, overcome constraints over transmission capacity, obtain regulatory incentives, and utilize the complete benefit of smart grid infrastructure.

The power quality and grid voltage can be improved and using distributed generation. Integration of RERs in microgrids helps in reducing the hazardous effects of GHG emission on the environment. Shared energy sources enable reduction of dependency on fuels and increase in energy security [30]. Smart homes are driven toward having their own uninterrupted renewable energy system that can operate on both stand-alone as well as grid-connected mode. Resynchronization algorithm and advanced islanding detection schemes can be used for uninterrupted and smooth transition between the modes. Plug-in hybrid electric vehicles (PHEV) can use the fuel cell output power for charging. The active and reactive power are controlled by the bidirectional power converter. The operating parameters of RERs vary based on the changing and unpredictable weather conditions and other parameters.

The frequency and voltage factors are crucial in grid control as the fluctuations in these parameters on transmission lines must be addressed by the grid operators within seconds to minutes. If any of these parameters are left unattended, the grid equipment and system may be damaged [18]. Reserve generators must be used for delivering power to the consumers within 10 minutes in case of occurrence of fault. Reactive power generators may be used for boosting up fall in voltage level in the main grid.

11.5 Microgrid Optimization Schemes

Smart microgrids are characterized by features like safe usage and power delivery, reliability, efficiently, distributed computing, advanced sensors, robust two directional communication, eco-friendliness, asset utilization optimization, resistance to cyber and physical attacks, consumer friendliness, and self-healing capabilities. AC, DC, or hybrid microgrid architecture can be employed for optimization of the gird. Energy storage devices, distributed generations, and mixed AC and DC loads are connected with an AC bus in AC microgrids. As the grid and most loads are of AC type, integration of AC microgrids into conventional AC grids is easier [26]. The flexibility, controllability, and capacity are larger for AC microgrids. However, the efficiency decreases significantly while DC energy storage devices, sources and loads are connected via DC/AC inverter to the AC bus.

A DC microgrid connects the AC/DC converter and the grid using a common DC bus. The DC microgrid's operation principles are similar to that of the AC microgrid. However, the AC and DC microgrid has similar operational principles. The power conversion losses are reduced in DC

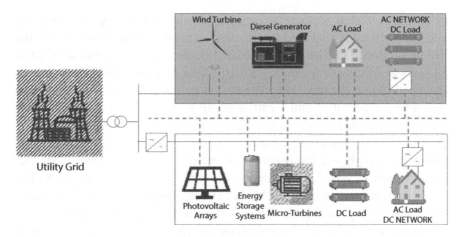

Figure 11.4 Hybrid AC/DC microgrid configuration.

microgrids when compared to AC microgrids as DC buses are sufficient for performing power conversion. The system size and cost are lesser and system efficiency is higher for DC microgrid [5]. The absence of reactive power offers improved stability and integration of DERs offer better compatibility in DC microgrids. Homopolar, bipolar, and monopolar are some of the several types of DC microgrids.

Both AC and DC loads, energy storage systems, and distributed generations are integrated directly in the same distribution grid, combining AC as well as DC microgrids, called hybrid AC/DC microgrid. Figure 11.4 represents the typical Hybrid AC/DC microgrid configuration. The black like represents the electrical network and red line represents the control and monitoring communication network. AC and DC loads, micro turbines, diesel generators, PV arrays, wind turbines, and energy storage systems are incorporated in this system [19]. They are connected via pertinent converters and interfaced with AC/DC buses. High reliability, lesser total cost, energy loss and conversion stages, easy DER integration, and minimal interface elements are the advantages of combining AC and DC microgrids into hybrid configuration. The storage and generation unit synchronization are not required in these systems as direct connection is established between the AC or DC network along with ESS, loads, and DG.

11.5.1   Load Forecasting Schemes

Load or demand forecasting is significant component in smart grids. During the specified time period, the future energy requirements of the system is predicted accurately through demand forecasting. The demand and generation are matched by the unit commitment strategies based on this prediction [31]. The planning can be performed for a duration of 20 to 50 years and prediction can be performed for on hourly basis of the operation process for the next 24 to 48 hours depending on the demands based on customer activities and weather conditions. The demand forecasting schemes are classified into two major categories, namely, artificial intelligence (AI)–based and statistical-based techniques [15]. Some of the commonly used AI-based forecasting models are fuzzy logic, particle swarm optimization (PSO), Least Squares–Support Vector Machine (LS-SVM) Algorithm, support vector machines (SVMs), improved variable learning rate back propagation (IVL-BP), Grey-Back Propagation (GBP) Neural Network, and Artificial Neural Network (ANN). The statistical-based schemes include Auto Regressive Integrated Moving Average (ARIMA), Auto Regressive Moving Average (ARMA), Moving Average (MA), and Auto Regressive (AR) models.

11.5.2   Generation Unit Control

Huge quantity of diverse DERs like fuel cells and wind or solar power generators are connected to the microgrid through power electronic interfaces or direct connection. The grid frequency and voltage are adjusted based on the information provided by the voltage source inverter (VSI). Grid-following and grid-forming are the two classifications of VSI. Voltage control is achieved between the loads and DG units by grid forming controller. The power obtained from DGs can be maximized in current-control mode by the grid-following controller. Stationary reference frame ($\alpha\beta$) and synchronous reference frame (dq) are grid-following schemes which use control strategies for DG units. The DG system stability may be improved by passivity-based control technique [4]. Various grid-following controllers are investigated for unbalance mitigation using symmetrical component transformation. The active and reactive power of DGs have to be regulated along with current and voltage control. The P/V and Q/f droop controllers are the commonly used techniques for droop control and reactive power compensation, respectively, in smart grids.

11.5.3   Storage Unit Control

Power balance between the load and renewable energy resources is established by the energy storage units in microgrids. In order to achieve this objective, appropriate charging and discharging control strategies are essential. Research solutions are provided for improving the output power of wind in wind farm energy storage systems using fuzzy control. Other strategies used for this purpose include Monte Carlo simulation, H-infinity control, sliding mode control, PI and PID control, neural network, and hysteresis current control [16–19].

11.5.4   Data Monitoring and Transmission

Smart grids are provided with bidirectional communication through advanced or smart metering infrastructure (AMI/SMI). The necessary software, hardware, communication system, and smart meters are integrated in the SMI for data procurement, storage, analysis, and energy utilization between the customer or utility and smart meter. At predefined intervals of time, recording and storage of real time energy consumption is performed by smart meters. Data transfer and communication can be done bidirectionally between the meter data management system (MDMS) and smart meters [3]. Affordability and acceptability by consumers play a major role in analysis of utilization of smart meters. Figure 11.5 provides the basic architecture of microgrid communication infrastructure.

Data is gathered from multiple neighborhood area networks (NAN) and transmitted to the control center for communication utility and smart grid through wide area network (WAN). The smaller area networks that are highly distributed and serve power systems at various locations are

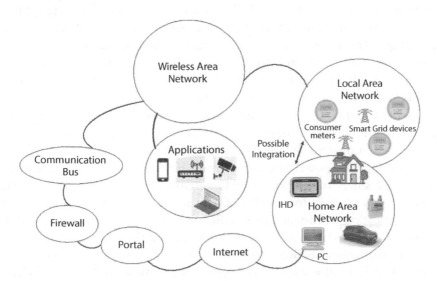

Figure 11.5 Architecture of microgrid communication infrastructure.

connected through WAN. Core and backhaul networks are the two major categories of WAN. PLC, 4G, WiMAX, and other similar technologies can also be used in the WAN networks. WAN enables data transfer at 10–100 Mbps speed over a distance of 1,000 sq. miles. The local/home area network (LAN/HAN) can be used in distributed energy generation facilities, in-home displays (IHD), programmable communicating thermostats, plug-in electric vehicles (PEVs), and other in-home smart appliances and devices. They support 10–100 kbps data transfer over a distance of 200 sq. meters. The power consumption is measured and the real time energy price is displayed to the consumers using IHD of HAN. The customers can minimize their electricity bill by customizing their power utilization profile based on the data provided by IHD [7–12].

In the distribution domain and customer domain, field gateways and smart meters are used, respectively, for establishing the communication network infrastructure in NAN. The energy data is collected from the smart meters using NAN and transmitted for control of other components. 4G, 3G, WiMAX, and other wired or wireless communication technologies can be used in this network [8]. The rate of data transfer is around 10 to 1,000 Kbps over long distances of 1 to 10 sq. miles. The AMI data is imported, verified, edited, and processed using MDMS before analysis and billing.

11.5.4.1   Communication Systems

Implementation of smart microgrids is possible due to the integration of communication technologies. The key deciding factors in choosing appropriate communication technology in microgrids include least repetitions, power quality, bandwidth, improved security features, and good transmission range. Both wired and wireless technologies can be incorporated in microgrids. Digital subscriber lines (DSL), optical communication, and power line communication (PLC) are the commonly used wired technologies used. Bidirectional communication is provided by integrating the power line cable with a modulated career in PLC. It is categorized into broadband and narrowband PLC [18]. Data as well as energy transfer is done through the power line cable. Power lines connect the data concentrator and smart meters in the PLC network. Further, cellular network technologies are used for transfer of data to the data center. The available communication infrastructure is utilized by the PLC, leading to lesser installation cost when compared to other communication systems.

Communication between control centers and substations is established widely by the use of fiber optic communication schemes. When compared to the traditional communication system, fiber optics offer less expensive, low loss, and high data transmission rate over long distances [16]. Telephone line wires are used for high-speed DSL communication. Installation cost is lesser as available communication infrastructure is used in this technology. Widespread availability, high data rate and low cost are the major advantages of DSL.

Wireless Sensor Network (WSN) technology does not involve the construction of complex infrastructure and can offer strong flexibility, self-healing power grid and high reliability. These features makes this technology a crucial component of smart grids. The network efficiency and stability can be improved by WSN. Real time maintenance, control, monitoring and exchange of bidirectional data is performed by WSN over the smart grid. Essential data is gathered from the target area and processed while monitoring the control devices [3]. The Carrier-sense multiple access with collision avoidance (CSMA/CA) principle is used in providing flexible and simple access structure of the member of IEEE 802.11 standards family, Wi-Fi, which is commonly used in LAN and HAN. It can operate on low-cost radio interfaces and unlicensed frequency bands of 2.4 to 5 GHz.

IEEE 802.11g and IEEE 802.11b are the most popular standards of IEEE 802.11. For outdoor and indoor environment, data rate of 1Mbps and 11Mbps are supported respectively by IEEE 802.11b, while a maximum of 54 Mbps data rate is supported by IEEE 802.11g. The most recent IEEE 802.11n provisions up to 150 Mbps data transfer rate. In most smart grids, the internet is accessed by smart devices through Wi-Fi for energy utilization management. HAN commonly use the Wi-Fi technology [12]. IEEE 802.16 standard, the WiMAX (Worldwide Interoperability for Microwave Access), can transfer data at a speed of 70 Mbps over a distance of up to 48 km, while supporting several thousand users simultaneously. Large coverage area, lower installation cost, automatic network connectivity, high data rate, and reliability are the major advantages of WiMAX technology in Smart Grid Applications.

Global System for Mobile (GSM) and General Packet Radio Service (GPRS) technologies are used for long distance transmission of data and control signals for communication between the utilities and smart meters [20]. GSM offers secure communication while GPRS operates on GSM-based wireless packet networks. Satellite technologies are commonly used for remote monitoring in smart grids especially in geographically remote or rural areas where other modes of communication are unavailable. This technology is expensive. However, equipping smart grids with satellite communication may be possible with the recent advancements in satellite systems. IEEE 802.15.4 standard, Zigbee technology is used for reduced cost, low complexity, lesser data rate, and low power–based communication. Computer peripherals, healthcare, smart meter, remote control, security systems, home automation, and other applications make use of Zigbee technology [31].

11.5.5   Energy Management and Power Flow

Reduced fuel consumption and cost cutting in power grids can be done along with improving the sustainability, reliability, and stability of the system using a control tool called energy management system (EMS) for controlling the flow of power between the load, DER, and main grid. System resynchronization takes place when transition occurs between island mode and grid. *Centralized and decentralized control* enabled with several hierarchical control are the major categories of EMS [29]. Data is gathered from the DERs of microgrids through a centralized controller. This data is further adjusted by the control equipment with a control variable and transmitted to the central system. Despite the low redundancy and reliability, centralized controller is most suited for small-scale microgrids.

During maintenance of this controller, the whole system has to be shutdown. Communication issues are another major drawback of this control. However, efficient solution is provided by the centralized hierarchical control from monetary perspective. The architecture of hierarchical controller is based on the existing infrastructure and microgrid type [2–6]. Distribution management system (DMS), microgrid central controller (MGCC), and local controllers (LC) are the three layers of the centralized hierarchical control scheme. The overall stability requirements and grid demands are met at DMS level. The DMS is integrated with each microgrid via MGCC. The storage requirements, load demand, and active power of the DER are determined by the MGCC for enabling the microgrid power management. Utility requirements are met by the two way communication between MGCC and LC. For the purpose of reliability, communication systems are sidestepped and the microgrid frequency as well as voltages are controlled using the local measurements by the local controllers.

The DER loads and units can be controlled independently through the decentralized EMS.

It is best suited for microgrids with diverse operational environment and aims. A communication bus is used for connecting the local controllers in this system. The DG and household can exchange data through the bus. In distributed system, the optimal output power may be determined without the use of MGCC in local controllers. This structure leads to releasing the stress on the microgrid communication network and significant reduction of computational needs [19]. Decentralized EMSs like multi-agent system (MAS) approach improves the stability of microgrid while determining the operation point of each DG using AI-based tools like fuzzy logic and neural network. It is intended to simplify complex and large systems into autonomous subsystems.

When compared to centralized EMS, the decentralized MAS is advantageous in several aspects. The computational time is reduces as the vital information from local controller is used and DGs are operated autonomously by MAS. Substantial data flow to the central node is essential for centralized control. It also offers plug and play capability. The utility grid flexibility, stability, reliability, and efficiency can be improved by the implementation of vehicle-to-grid (V2G) technologies in smart grids [28]. It allows power absorption from V2G to the utility grid or vice versa while charging the electric vehicle enabling bidirectional applications. Harmonic filtering, load balancing, active and reactive power compensation, electrical demand side management, spinning reserve, and frequency and voltage regulation are some of the benefits of this technology. The energy produced by RERs can also be stored in the electric vehicles [6].

11.6   Challenges in Implementation of Microgrids

Microgrids face challenges from the consumer, regional variation in power regulations, and other technical difficulties. The future smart grid end users that can contribute significantly toward the demand and supply load balancing issues are called smart consumers. Ease of use, power availability, maintaining comfort level, and decreasing electricity bill are the major requirements of smart consumers [10]. By using the ICT tools and smart data, consumers can participate actively toward demand management right from the level of domestic environment. The desired comfort level of the user is retained while supply constraints are met and cost is reduced by optimization of energy consumption through EMSs in smart homes.

The connection of main network, integration and penetration of DER, and facilitating microgrid applications are the major challenges in regulation. The limited regulations prevent the utilization of microgrids in their full potential [13]. The safety and functionality of main grid is to be preserved while interconnecting the main grid and microgrid. The grid has to disconnect immediately in case of blackouts or any faults. The high connection fee policies lead to high connectivity costs while integrating microgrids and main grids. The technical challenges include operation, compatibility of components, integration of renewable generation, and security [18]. The ability of microgrids to transition between islanded and grid connected mode leads to severe voltage and frequency control issues between the load and generation when there are huge mismatches. The plug and play feature may lead to challenges when large number of generation units are involved in the connect and disconnect operations.

Figure 11.6 provides the major barriers in the deployment of renewable energy technology. Control software, communication system, inverters, energy storage devices, CHP, fuel cell, microturbine, diesel generator, and several other components exist in a microgrid [15]. However, the characteristics of these components vary in terms of communication limits, control, rate of charging and discharging, energy storage, operation efficiency, operation cost, startup and shut down time, as well as power generation capacity. While integrating RES to main grid, weather dependence, unpredictability, and variability-related challenges are faced by the system. The stability of microgrid may be threatened by frequent and abrupt variations in the output power of RES. The distribution networks may also be congested with the increase in renewable shares. The scarcity of dispatch facility may lead to renewable energy generation intermittency [31].

Design of a security system that can respond to the faults of the microgrid as well as the main grid is challenging. While switching between the autonomous and grid-connected modes of operation, the magnitude of fault currents also varies significantly. The unidirectional flow of fault current in traditional power systems is replaced with bidirectional flow with the integration of microgrids. A fast static switch is used for interfacing the main power system with the microgrid for fault protection.

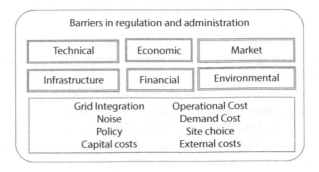

Figure 11.6 RET deployment barriers.

11.7   Future Prospects of Microgrids

Smart hybrid automatic transfer switch (HATS) and islanding detection method (IDM) can be integrated for managing the grid status and identifying the operating mode of microgrids. Unified power flow controller (UPFC), static series synchronous compensator (SSSC), static VAR compensator (SVC), and static synchronous compensator (STATCOM) use Flexible AC Transmission System (FACT) devices for solving issues with respect to reliability and stability while integrating RERs [32]. These devices are integrated with filters that mitigate harmonics caused by the power circuits. Further, stability of microgrid can be improved by various classification and analysis schemes.

The directional elements and overcurrent relays of the microprocessor are used for developing protection schemes under different modes of operations of microgrids. Differential protection, symmetrical component theory and adaptive protection system are some of the commonly used protection schemes. External protection devices like energy storage devices, fault current limiter and fast static switch may also be used. During transition of loads between islanded and grid-connected mode, high-speed isolation is provided by fast static switch. Operation cost and investment of microgrids, transmission lines, and candid generation units, along with total system cost can be minimized by using mixed integer programming optimization scheme. Cost function concept is used for minimizing power losses using high level controller while terminal voltage and power output can be regulated using low control layer based on decentralized voltage control algorithm.

11.8   Conclusion

Several challenges with respect to reliability and efficiency are faced by the power system with the growing global demands. Limited supply of primary energy has driven toward the utilization of distributed energy sources. Supply of improved power quality, need for storage units, lesser intermittency, and overcoming the glitches of uncertain power generation are some of the features that smart microgrids can provide while meeting the growing power demand. The ability to connect or disconnect from the major grid whenever necessary offers flexibility to the microgrid. The distribution network power losses may be reduced while offering improved power quality, reduced emission, lesser investment cost, and better reliability through microgrids. Smart grid functions and microgrid elements like monitoring schemes, data transmission, forecasting methods, and generation control are reviewed. Factors like power system stability, energy efficiency, demand side management, carbon dioxide emission, computer tools, and economic assessment are discussed along with the issues like consumer barriers and regulations. Future prospects of microgrids and opportunities to overcome the existing challenges are presented.

References

1. Badal, F.R., Das, P., Sarker, S.K., Das, S.K., A survey on control issues in renewable energy integration and microgrid. *Prot. Control Mod. Power Syst.*, 4, 1, 8, 2019.
2. Khalid, M., Ahmadi, A., Savkin, A.V., Agelidis, V.G., Minimizing the energy cost for microgrids integrated with renewable energy resources and conventional generation using controlled battery energy storage. *Renewable Energy*, 97, 646–655, 2016.
3. Hirsch, A., Parag, Y., Guerrero, J., Microgrids: A review of technologies, key drivers, and outstanding issues. *Renewable Sustainable Energy Rev.*, 90, 402–411, 2018.
4. Ourahou, M., Ayrir, W., Hassouni, B.E., Haddi, A., Review on smart grid control and reliability in presence of renewable energies: challenges and prospects. *Math. Comput. Simul.*, 167, 19–31, 2020.
5. Mohammed, Y.S., Mustafa, M.W., Bashir, N., Ibrahem, I.S., Existing and recommended renewable and sustainable energy development in Nigeria based on autonomous energy and microgrid technologies. *Renewable Sustainable Energy Rev.*, 75, 820–838, 2017.
6. Dawood, F., Shafiullah, G.M., Anda, M., Stand-alone microgrid with 100% renewable energy: A case study with hybrid solar PV-battery-hydrogen. *Sustainability*, 12, 5, 2047, 2020.
7. Yoldaş, Y., Önen, A., Muyeen, S.M., Vasilakos, A.V., Alan, İ., Enhancing smart grid with microgrids: Challenges and opportunities. *Renewable Sustainable Energy Rev.*, 72, 205–214, 2017.
8. Shoeb, M.A. and Shafiullah, G.M., Renewable energy integrated islanded microgrid for sustainable irrigation—A Bangladesh perspective. *Energies*, 11, 5, 1283, 2018.
9. Dobakhshari, A.S., Azizi, S., Ranjbar, A.M., Control of microgrids: Aspects and prospects, in: *2011 International Conference on Networking, Sensing and Control*, 2011, April, IEEE, pp. 38–43.
10. Xu, C. and Lu, W., Development of smart microgrid powered by renewable energy in China: current status and challenges. *Technol. Anal. Strateg. Manage.*, 31, 5, 563–578, 2019.
11. Coelho, V.N., Cohen, M.W., Coelho, I.M., Liu, N., Guimarães, F.G., Multi-agent systems applied for energy systems integration: State-of-the-art applications and trends in microgrids. *Appl. Energy*, 187, 820–832, 2017.

12. Ullah, S., Haidar, A.M., Hoole, P., Zen, H., Ahfock, T., The current state of Distributed Renewable Generation, challenges of interconnection and opportunities for energy conversion based DC microgrids. *J. Cleaner Prod.*, 273, 122777, 2020.

13. Kiptoo, M.K., Lotfy, M.E., Adewuyi, O.B., Conteh, A., Howlader, A.M., Senjyu, T., Integrated approach for optimal techno-economic planning for high renewable energy-based isolated microgrid considering cost of energy storage and demand response strategies. *Energy Convers. Manage.*, 215, 112917, 2020.

14. Mahfuz, M.U., Nasif, A.O., Hossain, M.M., Rahman, M.A., Integration of Renewable Energy Resources in the Smart Grid: Opportunities and Challenges, in: *Transportation and Power Grid in Smart Cities: Communication Networks and Services*, pp. 291–325, 2018.

15. Yang, Y., Fang, Z., Zeng, F., Liu, P., Research and prospect of virtual microgrids based on energy internet, in: *2017 IEEE Conference on Energy Internet and Energy System Integration (EI2)*, 2017, November, IEEE, pp. 1–5.

16. Samad, T., Koch, E., Stluka, P., Automated demand response for smart buildings and microgrids: The state of the practice and research challenges. *Proc. IEEE*, 104, 4, 726–744, 2016.

17. Lehtola, T. and Zahedi, A., Solar energy and wind power supply supported by storage technology: A review. *Sustain. Energy Technol. Assess.*, 35, 25–31, 2019.

18. Vadi, S., Padmanaban, S., Bayindir, R., Blaabjerg, F., Mihet-Popa, L., A review on optimization and control methods used to provide transient stability in microgrids. *Energies*, 12, 18, 3582, 2019.

19. Hussain, S.M., Nadeem, F., Aftab, M.A., Ali, I., Ustun, T.S., The emerging energy internet: Architecture, benefits, challenges, and future prospects. *Electronics*, 8, 9, 1037, 2019.

20. Soshinskaya, M., Crijns-Graus, W.H., Guerrero, J.M., Vasquez, J.C., Microgrids: Experiences, barriers and success factors. *Renewable Sustainable Energy Rev.*, 40, 659–672, 2014.

21. Liu, X. and Su, B., Microgrids—an integration of renewable energy technologies, in: *2008 China International Conference on Electricity Distribution*, 2008, December, IEEE, pp. 1–7.

22. Adefarati, T. and Bansal, R.C., Reliability and economic assessment of a microgrid power system with the integration of renewable energy resources. *Appl. Energy*, 206, 911–933, 2017.

23. Ahmed, M., Amin, U., Aftab, S., Ahmed, Z., Integration of renewable energy resources in microgrid. *Energy Power Eng.*, 7, 01, 12, 2015.

24. Lv, T. and Ai, Q., Interactive energy management of networked microgrids-based active distribution system considering large-scale integration of renewable energy resources. *Appl. Energy*, 163, 408–422, 2016.

25. Tooryan, F., HassanzadehFard, H., Collins, E.R., Jin, S., Ramezani, B., Smart integration of renewable energy resources, electrical, and thermal energy storage in microgrid applications. *Energy*, 212, 118716, 2020.

26. Montuori, L., Alcázar-Ortega, M., Álvarez-Bel, C., Domijan, A., Integration of renewable energy in microgrids coordinated with demand response resources: Economic evaluation of a biomass gasification plant by Homer Simulator. *Appl. Energy*, 132, 15–22, 2014.
27. Wilson, D.G., Robinett, R.D., Goldsmith, S.Y., Renewable energy micro-grid control with energy storage integration, in: *International Symposium on Power Electronics Power Electronics, Electrical Drives, Automation and Motion*, 2012, June, IEEE, pp. 158–163.
28. Basak, P., Chowdhury, S., nee Dey, S.H., Chowdhury, S.P., A literature review on integration of distributed energy resources in the perspective of control, protection and stability of microgrid. *Renewable Sustainable Energy Rev.*, 16, 8, 5545–5556, 2012.
29. Al-Ghussain, L., Samu, R., Taylan, O., Fahrioglu, M., Sizing renewable energy systems with energy storage systems in microgrids for maximum cost-efficient utilization of renewable energy resources. *Sustain. Cities Soc.*, 55, 102059, 2020.
30. Falahati, B., Kargarian, A., Fu, Y., Timeframe capacity factor reliability model for isolated microgrids with renewable energy resources, in: *2012 IEEE Power and Energy Society General Meeting*, 2012, July, IEEE, pp. 1–8.
31. Zia, M.F., Elbouchikhi, E., Benbouzid, M., Microgrids energy management systems: A critical review on methods, solutions, and prospects. *Appl. Energy*, 222, 1033–1055, 2018.
32. Fouladi, E., Baghaee, H.R., Bagheri, M., Gharehpetian, G.B., Power Management of Microgrids including PHEVs based on Maximum Employment of Renewable Energy Resources. *IEEE Trans. Ind. Appl.*, 56, 5, 5299–5307, 2020.

12

Challenges in Planning and Operation of Large-Scale Renewable Energy Resources Such as Solar and Wind

J. Vishnupriyan[1] and A. Dhanasekaran[2*]

[1]*Center for Energy Research, Department of Electrical and Electronics Engineering, Chennai Institute of Technology, Chennai-600069, India*
[2]*Center for Energy Research, Department of Mechanical Engineering, Chennai Institute of Technology, Chennai-600069, India*

Abstract

Energy is one of the major inputs for the economic development of any country. Since the industrial revolution, the energy production of most countries across the world has been dominated by fossil fuels. The by-products of fossil fuels have created major implications in the global climate as well as in the human health. Three-quarters of global greenhouse gas emissions have resulted from the burning of fossil fuels to produce energy. To reduce CO_2 emissions and pollution, the world needs a rapid shift towards low-carbon sources of energy like renewable energy technologies. The renewable energy sources play a key role in the decarbonization of energy systems. Many countries have installed significant amount of solar and wind power plants to meet their energy demand. Among the various types of renewable energy sources, most of the energy come from solar and wind power. Around 724.09 TWh energy was generated by solar power and 1429.62 TWh energy was generated by using wind energy. Nevertheless, solar and wind tend to be more variable and uncertain than its counterpart, the planning and operation of large-scale renewable energy resources such as solar and wind will remain a challenge that needs a thorough understanding in the power system operation and planning strategies. This chapter discusses the main challenges which are associated in planning and operation of large-scale renewable energy resources such as solar and wind when they are integrated with grid.

Corresponding author: dhanasekarana@citchennai.net

M. Kathiresh A. Mahaboob Subahani and G.R. Kanagachidambaresan (eds.) Integration of Renewable Energy Sources with Smart Grid, (263–280) © 2021 Scrivener Publishing LLC

Keywords: Frequency control, grid integration, power quality, renewable energy, solar PV system, voltage control, wind energy system

12.1   Introduction

The socio-economic development of any country depends on its energy production and utilization. In all the countries, the most of energy is produced by means of fossil fuels. However, consumption of such fossil fuels will lead to air pollution, climate change, depletion of the ozone layer, and acid rain. Besides, the production of greenhouse gases and depletion of the natural resources has caused the worldwide concern. One alternative, to meet the large energy demand is employing nuclear power plants due to its large power generating capacity. Since nuclear energy does not produce any greenhouse gases, it can be considered as an alternative to the fossil fuel energy generation as well as a solution for mitigating the pollution. However, nuclear power plants create a wide variety of environmental and health issues. Moreover, the generation of radioactive wastes is one of major problem in nuclear power plants which can remain radioactive and hazardous to all living creatures for thousands of years. Thus, the renewable energy is the only available alternative source to generate sustainable power in future for all countries.

The average energy received by the Earth's surface from the sun is 1.2×10^{17} W, less than an hour of this power can meet the energy demand of the whole population for a year [1]. Solar and wind are the most potential renewable energy sources capable to produce enormous quantities of power. The utilization of sustainable energy will lead to decarbonisation, energy security, and improvement in the energy access. Today, renewable energy resources are integrated with the existing electrical energy sources to achieve clean and sustainable power generation.

Grid integration is a technique to deliver variable renewable energy to the grid for its effective utilisation. In recent years, robust integration methods have been adopted to maximize the cost-effectiveness, system stability, and reliability of variable renewable energy sources when it is integrated into the power system. The large-scale renewable power generation can be connected to transmission systems and the small-scale power generation can be connected to distribution systems. However, there are certain challenges in the integration of renewable energy to the existing grid due to intermittent nature of renewable energy sources. The uncertainty in the wind speed makes it difficult to obtain good quality power. The fluctuations in the wind speed affect the voltage and active power output of the electric machines which are connected to the wind turbine. Similarly, solar photovoltaic (PV) energy penetration alters the voltage profile and frequency response of the system and influences the transmission and distribution systems [2]. The renewable energy resources which are intermittent in nature may be predictable but cannot be dispatched or transmitted in a flexible way to meet the fluctuating demand since the output of renewable energy sources cannot be altered rapidly. This chapter discusses the challenges in planning and operation of large-scale renewable energy resources such as solar and wind.

12.2   Solar Grid Integration

Solar grid integration is a network which allows substantial penetration of PV power into the utility grid. The integration of standardized PV systems into grids help to optimize the energy balance, decrease operating cost, and provide added value to the consumer and the utility. The solar power generation used to integrate with grid are Concentrated Solar Power (CSP) and Solar Photovoltaic (SPV) power.

The CSP generation is similar to that of a conventional thermal power plant (i.e., converting thermal energy into electrical energy). Whereas, the PV power systems use sunlight to generate Direct Current (DC). It is then converted to Alternating Current (AC) with the help of inverters and other components before it is distributed into the power grid. Since PV systems directly generate electricity, it cannot be easily stored in batteries at large power levels. However, CSP systems can store the energy using thermal energy storage technologies. Since energy storage technology can help to overcome intermittency problems, CSP can attain better penetration in the power generation industry. Nevertheless, the technology and the high initial cost are currently limiting the CPS for its large-scale expansion and deployment. Therefore, energy market conditions are currently favour the PV energy installations.

Solar grid integration allows large-scale solar power produced from PV system to penetrate the already existing power grid. It requires careful considerations and attentions at various levels from design and manufacturing of PV systems, installation, and operation. The solar energy penetration into the transmission grid requires a thorough understanding of the effects on the grid at various points. A typical PV plant that feed electrical energy into the grid consists of the following components, PV generator, inverter, transformers, generator junction box, meters, grid connection, and DC and AC cabling and its accessories. Among the various components, the inverters play a major role in a PV system as it converts the DC output into AC which is the main requirement of all commercial appliances. The inverters have to supply constant voltage and frequency irrespective of varying load conditions and need to supply or absorb reactive power in the case of reactive loads. Further, they reconcile the systems with each other to feed the solar power into the grid with maximum possible efficiency. Therefore, the effectiveness of PV installations strongly dependent on the reliability and efficiency of the inverters [3]. Hence, inverters can be considered as the brain of a PV grid power systems. Generally, PV power systems are classified as very large-scale (higher than 100 MW), large-scale (1–100 MW), medium-scale (250–1,000 kW), and small-scale (up to 250 kW) PV power system [4]. The typical layout of a PV power system integrated with grid is shown in Figure 12.1.

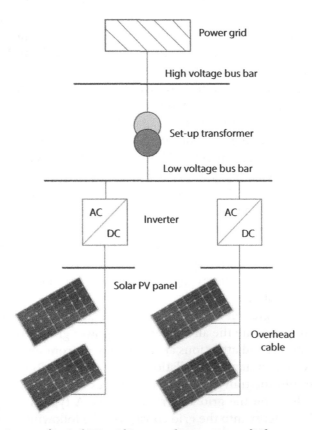

Figure 12.1 Layout of typical PV-grid integrated power system [10].

The connection of various elements depends on the network topology used by the inverter. Two commonly used topologies to connect the PV systems with grid are central and multistring inverter.

12.3   Wind Energy Grid Integration

Wind is an inexhaustible type of energy. Wind energy is clean and freely available. It originates from the non-uniform of heating of the atmosphere by the sun's rays, the surface irregularities of the earth, and its rotation. The wind energy can be converted to electricity with the help of wind turbines which convert the kinetic energy of wind into mechanical energy which in turn run the generator to produce electricity. The electrical energy is then fed into grid. The integration of electrical energy produced from wind to the grid can be made at low-voltage (nominal voltage up to 1 kV), medium-voltage (nominal 1 kV up to 35 kV), and high-voltage (nominal voltage above 35 kV), as well as extra high-voltage system.

The main components of the wind turbines are turbine rotor, gear box, and generator. The rotor converts the wind energy into mechanical energy. The generator converts the mechanical energy into electrical energy [5]. The gear box is used to adapt the rotor speed to the generator speed if necessary. Figure 12.2 shows the system overview of wind energy grid integration system.

Large and heavy industry can be connected to the high voltage system, medium size industries and workshops can be connected to the medium voltage system, while small consumers can be connected to the low voltage system. The major components of the grid for connection of the wind turbine systems are transformer, substation with safety equipment, and electricity meter. Since the losses are high in low voltage lines, wind turbines have its own transformer to convert the voltage level of the turbine to the medium voltage of the distribution system.

12.4   Challenges in the Integration of Renewable Energy Systems with Grid

The intermittent nature of solar and wind energies bring some distinct characteristics in the power generation. They not only impact the operation and security of the grid but also create new challenges to the owners of the renewable energy generation and the grid operators. The main characteristics of renewable energy that affect power quality are non-controllable variability, partial unpredictability, and location dependence. The power

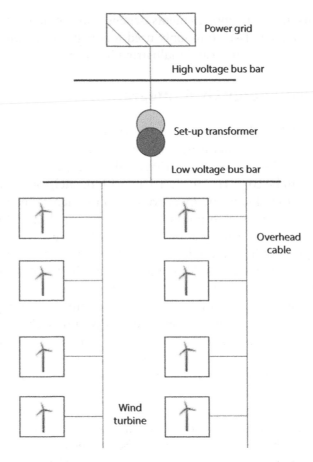

Figure 12.2 Overview of wind energy grid integration system [10].

quality issues in the renewable energy integrated grid arise mainly due to the following factors:

1. Non-controllable variability of solar and wind energy resources;
2. Due to the disturbances which are arising in the power grid;
3. Harmonics produced by power electronic devices.

As per the guidelines of IEEE standard 929-2000, voltage, voltage flicker, frequency, and distortion are some of the major parameters used to evaluate the quality of power produced in the renewable energy systems. Any variation from the recommended value is the indication of reduction in the power quality. Some of the characteristics of renewable energy systems which influences of power quality are discussed in the following sections.

a. Non-controllable variability

The wind and solar output are varying in nature so that operators at the generation plant cannot control them. Since wind speed and available sunlight may vary from time to time, the output power also vary accordingly. These fluctuations in the output power not only requires additional energy on an instantaneous basis to balance supply and demand but also requires ancillary services such as frequency regulation and voltage support.

b. Partial unpredictability

The wind and sunlight availability are partially unpredictable. A wind turbine may generate electricity only when the wind is blowing at a certain velocity. Similarly, solar PV systems require the presence of sunlight to produce energy. The partial unpredictability can be minimised by adopting improved forecasting technologies and effective maintenance of reserves which provide additional power.

c. Location dependence

The solar and wind resources are location specific. Unlike fossil fuels the renewable resources cannot be transported easily. This indicates that renewable energy generation must be at the same location where the resources are available. Often, these locations are at a long distance from the places where the utility is located. Hence, new transmission capacity is to be employed to integrate renewable energy sources to the grid. Further, transmission costs are also important especially for off-shore wind energy resources due to the fact that they require the use of specific technologies which are not used in land-based transmission lines. However, local unpredictability can be improved with the help of advanced forecasting methods which are primarily used to predict the weather and energy output from the renewable energy resources accurately at several timescales [6].

12.4.1   Disturbances in the Grid Side

The Power grid side disturbances are one of major problem which affect renewable energy integration. The power grid-code requirements were developed in several countries to attain reliable power system operation in the grid-integrated renewable energy systems. Solar power plants have to undergo certain disturbances without disconnecting from the grid [7]. The Institute of Electrical and Electronics Engineers standard (IEEE1547a) has released a new definition for the voltage sag trip settings to enable the equipment to ride through during voltage sag. This new definition does not allow the distributed generation to trip when the duration of voltage sags is between default settings and maximum settings, and help to arrive an agreement between the distributed resource owner and the local utility. The control systems of the inverter and the energy storage can be accordingly designed in order to improve the PV power plants ride through capability so that proper voltage level can be maintained during voltage sags caused by power grid side disturbances.

a. Voltage and frequency fluctuations

In a renewable energy and grid integrated power system, voltage variation can be related to reactive power, whereas frequency variation can be determined from the changes in the real power. Effective control of reactive power and real power enables the smoothening of voltage and frequency.

b. Voltage control

The power quality disturbances of renewable energy systems are mainly due to change in load and in the environmental effects. The effective operation of a PV system depends upon various factors such as solar intensity, cell temperature, shading effect, and the semiconductor devices. A similar condition is also existing even in the operation of a wind turbine with any configuration used in design and operation. The power fluctuation of a wind turbine is influenced by various factors like type of turbine, controlling algorithm, wind velocity, and the shadow effect of the tower which lead to fluctuations in the power density and power quality disturbances like voltage sag-swell along with power flow [8]. The SPV systems integrated to the grid do not regulate voltage but it will inject current into the grid. It may be noted that the operating range for the voltage in the inverters is selected for power system protection not for voltage regulation. More current injection into the power grid will affect the voltage in the utility. If the PV system injects less current than the load on that line, the voltage regulator of the utility will operate normally. On the contrary, if the current injection is more than that of the load, then the corrective measures are required due to the lack of bi-directional current sensing capabilities of voltage regulating devices. If the operating voltages are not specified for solar PV-grid systems in IEEE standards, then it can be set a value between 88% and 110 % of the nominal voltage [9].

Similarly, for a wind power plant power system, IEC Standard 61400-21 has recommended a 10-minute average of voltage fluctuation can be within ±5% of its nominal voltage [10]. A wide variation in voltage can cause errors in operation and performance deterioration. Therefore, it is essential to maintain the power at substantially constant voltage. The reactive power compensation is an effective method to control the voltage. The Flexible Alternating Current Transmission System (FACTS) devices can be employed to increase the power quality in the transmission lines.

The FACTS System consists of power electronic devices as well as power system devices to enhance the control and stability and to increase the capability of the transmission system. Various types of FACTS are as follows,

a. Series controllers

The series controllers have capacitors or reactors which introduce voltage in series with the line. These controllers are variable impedance devices and their primary function is to reduce the inductivity in the transmission line. Series controllers can supply or consume variable reactive power. Some of the series controllers are SSSC, TCSC, TSSC, etc.

b. Shunt controllers

The shunt controllers consist of variable impedance devices like capacitors or reactors to introduce current in series with the line. These controllers are used to reduce the capacitive of the transmission line. The injected current is in phase with the line voltage. Some of the commonly used shunt controllers are STATCOM, TSR, TSC, and SVC.

c. Shunt-Series controllers

The shunt-series Controllers introduce current in series using the series controllers and voltage in shunt using the shunt controllers. An example for shunt-series controller is UPFC.

d. Series-Series controllers

The series-series Controller is a combination of series controllers with each controller provides series compensation and also transfers real power along the line. One commonly used series-series controller is IPFC.

12.4.2   Virtual Synchronous Machine Method

Generally, the conventional power plants maintain and regulate voltage and frequency during disturbances due to the fact that the synchronous generator stores kinetic energy due to inertia. The absorbed kinetic energy is then released to compensate any imbalances that present in electrical and mechanical power in the generator [11]. The large masses of rotational elements in the generators can provide considerable amount of inertia in the power grid. Whenever a frequency variation occurs the inertial reserve of power system responds and balances the initial frequency variation and brings the frequency back to its steady state value prior to primary reserve [12]. On the contrary, the power electronic inverters employed in the renewable energy grid integrated systems have no rotational energy so the inertia of these inverters is negligibly small. Moreover, extensive usage of power electronic devices can cause a reduction in the equivalent rotational inertia of power grid and large frequency oscillation [13]. In order to overcome this problem and to improve the power system stability, the Virtual Synchronous Machine (VSM) method was proposed by Beck and Hesse in 2007.

A typical configuration VSM is shown in Figure 12.3. The VSM models grid integrated renewable energy generators as an electromechanical synchronous machine. The VSM combines the dynamic power electronic inverter technology and the operating characteristics of synchronous machines to control the grid-interface converter of a generator as well as an energy storage [14]. In VSM, renewable generation side has energy storage. The combined arrangement of renewable energy and Energy Storage System (ESS) can regulate voltage and operate as the stator output of VSM. The VSM can handle active power and reactive power flow in both directions and enhances power quality and grid stability for the distributed generation.

12.4.3   Frequency Control

The electric power systems have two important characteristics namely, voltage and frequency. In order to maintain the expected operating conditions and to supply the energy to the loads, it is necessary to control these two parameters within the limits to avoid unexpected disturbances which cause the power system to fail. The integration of renewable energy resources into the grid introduces a challenge in frequency control in the power systems. Frequency stability and its control is the major problem in the design and operation of integrated grid systems. The frequency control in an integrated power grid has to perform complex multi-objective regulation optimization

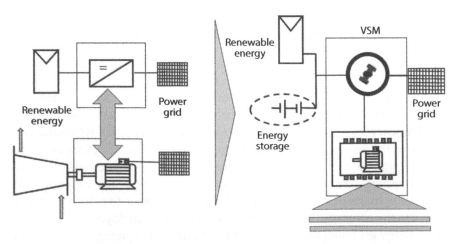

Figure 12.3 Configuration of virtual synchronous machine [10].

problems which are characterized by a high degree of diversification in management policies and in the demand and supply sources.

Frequency stability can be referred as the ability of a power system to keep the system frequency within the specified operating limits. The frequency instability is a result of imbalance between load and generation which is mainly associated with poor coordination of control and protection of equipment, insufficient generation reserves, and inadequacies in equipment responses. A deviation of frequency from its recommended operating range may affect power system operation, security, reliability, and efficiency by damaging equipment, degrading load performance, overloading transmission lines, and triggering the protection devices. Different types of frequency control loops are required to maintain power system frequency stability depending on the amplitude and duration of frequency deviation.

Various control techniques are required to be employed in the power system to improve the power quality from renewable energy sources and to reduce the undesirable disturbances in the power grid. The frequency control techniques applied to wind and solar energy systems are shown in Figure 12.4. The frequency can be controlled by reserving the active power

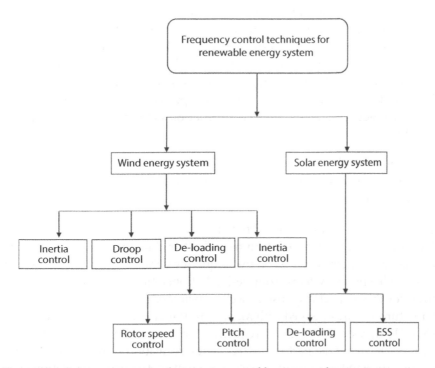

Figure 12.4 Frequency control techniques in renewable energy grid integration power systems.

by using either de-load operation or an ESS. Frequency control methods are classified as control techniques to de-load the renewable energy systems and control practices for renewable energy systems with ESS. The frequency controlling methods for wind energy generators are inertia control, droop control, and de-loading control.

a. Inertia control

The inertia control needs to be emulated to regulate the frequency in the high penetration wind generators. This can be done by hidden inertia emulation or by the fast power reserve emulation.

1. Hidden inertia emulation

The hidden inertia emulation in the wind energy generator is used to control the frequency variation and to maintain its stability. The control algorithm for the power electronic converters enables the wind turbine to deliver the kinetic energy and help to regulate the frequency through the inertial response [15].

2. Fast power reserve

The fast power reserve is the constant active power support irrespective of the wind speed and it is the temporary power which is released from the kinetic energy of the wind turbine generators.

The fast power reserve control operates when the deviation in the frequency is higher than the threshold value. It delivers the additional power in the frequency through the kinetic energy and is known as over-production.

The recovery of kinetic energy after the event is called under-production. In order to avoid sudden dip in the active power [16], the change from an over-production state to an under-production state has to be accomplished steadily.

b. Droop control

The droop control modifies the active power response according to its the frequency deviation.

c. De-loaded operation in wind turbines

De-loaded operation in wind turbines helps to reduce the adverse effect of the higher RES penetration level on the frequency stability. The reserve power in renewable energy systems from de-loaded operation can be effectively utilized for the inertial response as well as primary frequency support. Since, wind turbines are operated with Maximum Power Point Tracking Technique (MPPT) they do not regulate frequency. Hence, to maintain stability, wind turbines have to take part in the frequency regulation so de-loaded control enables the wind turbines to actively participate in frequency regulation [17].

The de-loading operation of the wind turbine can be realized by varying its operating point from the maximum to sub-optimal power point.

To regulate the active power output in the wind turbines, two methods are used: one is over-speeding and the other is under-speeding. In over-speed mode, the wind turbine delivers the kinetic energy till the operating point reaches its maximum value. In under-speed mode, it absorbs the kinetic energy till the operating point reaches the maximum value [18]. Since the under-speed control mechanism can cause stability problems, the over-speeding of the rotor control can be recommended generally.

12.4.4   Solar Photovoltaic Array in Frequency Regulation

The PV energy systems do not participate in active reserve power to regulate the frequency. All the PV systems are operating at its maximum power point using MPPT. In PV energy systems, the inertial and primary frequency response can be executed by using de-loaded operation of PV and ESS with suitable control algorithms.

12.4.5   Harmonics

The integration of renewable energy sources with grid can create harmonics due to the presence of power electronics devices as well as inverters which are connected to the renewable energy sources. Among the various important power quality challenges, the harmonics comes on top since they affect the voltage and current quality at the Point of Common Coupling (PCC) and negatively affects the loads. For example, PV systems are the most used renewable energy generators. It is connected to the low voltage distribution grid using power electronics devices and with the increased penetration level massive harmonic current will be injected into the network.

Maintaining adequate harmonics in the line currents of renewable grid integrated power system is one of the biggest challenges for the power system operators. Various harmonics mitigation and compensation methods have been developed and used in the power systems. The conventional method of reducing harmonics in power systems is the use of passive filters, active filters, filters in shunt orientation, and hybrid filter. Active power filters are commonly used to reduce harmonics. Some of the advanced control methods to compensate harmonics [19] are Current Control Method (CCM), the Voltage-Control Method (VCM), and the Hybrid-Control Method (HCM). The harmonic compensation can be done at local loads or at the point of common coupling.

The virtual impedance method is also increasingly utilized for controlling voltage-source and current source converters primarily driven by the renewable energy grid integration systems. For active stabilization and disturbance rejection, this method can be embedded as an additional degree of freedom. To increase system stability and to provide adequate distribution of harmonic compensation in the multiple distributed generation units the virtual impedance can be placed between the interfacing converter output and main power grid [20]. The key aspects for the virtual impedance method are the optimal design value, robust implementation, and proper utilization of the virtual impedance. The main objective of utilizing virtual impedance approach is to improve system performance during transient and to mitigate harmonics.

The harmonic compensation can also be done by using Multifunctional Grid-Tied Inverter (MFGTI) technique [21]. This is an advanced grid-tied inverter which is used not only to interface the renewable energy into power grid but also to enhance power quality at its grid-tied point. Since the capacity of a grid-tied inverter is higher than that of PV and wind energy systems, it can accommodate the intermittent features of solar and wind energy. These grid-tied inverters can be utilized to enhance the power quality so power quality conditioner may be required in an inverter-dominated power system.

12.5 Electrical Energy Storage (EES)

The electrical energy storage on a large scale is a major focus of attention at present due to the intermittent nature of renewable energy. The rapid increase of generating capacity by renewable energy sources has led to the development of energy storage on a large scale. Electricity cannot itself be stored but it can be converted to other forms of energy and later reconverted to electricity based on the demand. Some of the electrical energy storage systems are battery, flywheel, compressed air, and pumped hydro storage. Figure 12.5 shows the classification of various types of energy storage used in renewable energy power systems.

EES can be employed for the following reasons:

1. To reduce energy costs by storing electricity obtained at off-peak hours and using it at peak hours.
2. To help the customers by improving the reliability of the power supply and power quality during power grid failures [22].

The main requirement in power utilities is to ensuring the power quality. The fluctuation in the solar and wind energy output makes the system frequency control a difficult one and this can be overcome by employing EES. Further, EES can be used to provide voltage adjustment in power grid even

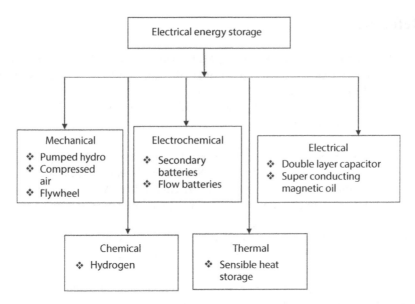

Figure 12.5 Classification of energy storages used in renewable energy power systems [10].

though the voltage is controlled by using transformer tap changers and reactive power with phase modifiers. By charging and discharging, the EES can control the voltage spikes. It can also be incorporated with advanced control systems to improve power system stability, reliability, and power quality in the renewable energy system and grid integration [23].

12.6 Conclusion

This chapter enumerated the challenges associated in the power quality due to integration of renewable energy systems with grid. The various methods to mitigate the voltage and frequency fluctuations and harmonics. Further, the electrical energy storage systems and its role in power quality improvement were also discussed. The future research and development directions of power quality improvement will be based on the following factors such as,

1. Employing VSM method for controlling the voltage and frequency.
2. The virtual impedance method for harmonic compensation.
3. Obtaining field measurements and information on harmonic spectrums for renewable energy systems to evaluate its characteristics.
4. New control schemes for the inverters can be introduced to attenuate the disturbances which arise at the grid.

References

1. Alshahrani, A., Omer, S., Su, Y., Mohamed, E., Alotaibi, S., The technical challenges facing the integration of small-scale and large-scale PV systems into the grid: a critical review. *Electronics*, 8, 1443, 2019.

2. Sandhu, M. and Thakur, T., Issues, challenges, causes, impacts and utilization of renewable energy sources - grid integration. *Int. J. Eng. Res. Appl.*, 4, 636–643, 2014.

3. Nwaigwe, K.N., Mutabilwa, P., Dintwa, E., An overview of solar power integration into electricity grids. *Mater. Sci. Energy Technol.*, 2, 629–633, 2019.

4. Rakhshani, E., Rouzbehi, K., Sánchez, A.J., Tobar, A.C., Pouresmaeil, E., Integration of large-scale PV-based generation into power systems: a survey. *Energies*, 12, 1425, 2019.

5. Schulz, D., Grid integration of wind energy systems, in: *Power electronics in smart electrical energy networks*, Springer, London, 2008.

6. IEC, *Grid integration of large-capacity renewable energy sources and use of large-capacity electrical energy storage*, International electrotechnical commission, pp. 1-104, Geneva, Switzerland, 2012.

7. Honrubia-Escribano, A., García-Sánchez, T., Gómez- Lázaro, E., Muljadi, E., Molina-García, A., Power quality surveys of photovoltaic power plants: characterisation and analysis of grid-code requirements. *IET Renewable Power Gener.*, 9, 5, 466–473, 2015.

8. Ray, P.K., Mohanty, S.R., Kishor, N., Classification of power quality disturbances due to environmental characteristics in distributed generation system. *IEEE Trans. Sustain. Energy*, 4, 2, 302–313, 2013.

9. IEEE Recommended practice for utility interface of photovoltaic systems (PV), in: IEEE Std 929-2000, 1-23, 2000.

10. Liang, X., Emerging Power Quality Challenges Due to Integration of Renewable Energy Sources. *IEEE Transactions on Industry Applications*, 53, 2, 855–866, 2017.

11. Ray, P.K., Mohanty, S.R., Kishor, N., Classification of power quality disturbances due to environmental characteristics in distributed generation system. *IEEE Trans. Sustain. Energy*, 4, 2, 302–313, 2013.

12. Tamrakar, U., Galipeau, D., Tonkoski, R., Tamrakar, I., Improving transient stability of photovoltaic- hydro microgrids using virtual synchronous machines. *IEEE Eindhoven PowerTech*, 2015, pp. 1–6, 2015.

13. Van Thong, V., Woyte, A., Albu, M., Van Hest, M., Bozelie, J., Diaz, J., Loix, T., Stanculescu, D., Visscher, K., Virtual synchronous generator: laboratory scale results and field demonstration. *2009 IEEE Bucharest power tech*, pp. 1–6, 2009.

14. Chen, Y., Hesse, R., Turschner, D., Beck, H.-P., Improving the grid power quality using virtual synchronous machines. *International conference on power engineering, energy and electrical drives*, pp. 1–6, 2011.

15. Knudsen, J.N.N.H., Introduction to the modelling of wind turbines, in: *Wind power in power systems*, Wiley, Chichester, UK, 2005.

16. El Itani, S., Member, S., Annakkage, U.D., Member, S., Joos, G., Short-term frequency support utilizing inertial response of DFIG wind turbines, in: *Proceedings of the 2011 IEEE Power and Energy Society General Meeting*, San Diego, CA, USA, 24–29, pp. 1–8, 2011.

17. Ekanayake, J., Holdsworth, L., Jenkins, N., Control of DFIG wind turbines. *Power Eng.*, 117, 28–32, 2003.

18. Ghosh, S., Member, S., Senroy, N., Electromechanical dynamics of controlled variable-speed wind turbines. *IEEE Syst. J.*, 9, 639–646, 2015.

19. Li, Y.W. and He, J., Distribution system harmonic compensation methods: an overview of DG-interfacing inverters. *IEEE Ind. Electron. Mag.*, 8, 4, 18–31, 2014.

20. He, J. and Li, Y.W., Analysis, design, and implementation of virtual impedance for power electronics interfaced distributed generation. *IEEE Trans. Ind. Appl.*, 47, 6, 2525–2538, 2011.

21. Zeng, Z., Yang, H., Tang, S., Zhao, R., Objective-oriented power quality compensation of multifunctional grid-tied inverters and its application in microgrids. *IEEE Trans. Power Electron.*, 30, 3, 1255–1265, 2015.

22. IEC, *Grid integration of large-capacity renewable energy sources and use of large-capacity electrical energy storage*, International electrotechnical commission, PP.1–104. Geneva, Switzerland, 2012.

23. Miguel A. Torres L., Luiz A. C. Lopes, Luis A. Morán T., and José R. Espinoza C., Self-Tuning Virtual Synchronous Machine: A Control Strategy for Energy Storage Systems to Support Dynamic Frequency Control, *IEEE Transactions on Energy Conversion*, 29, 4, 833–840, 2014.

13

Mitigating Measures to Address Challenges of Renewable Integration—Forecasting, Scheduling, Dispatch, Balancing, Monitoring, and Control

K. Latha Maheswari[1]*, B. Sathya[1] and A. Maideen Abdhulkader Jeylani[2]

[1]Department of Electrical and Electronics Engineering PSG College of Technology, Coimbatore, India
[2]Department of Electrical and Electronics Engineering, Sri Krishna College of Engineering and Technology, Coimbatore, India

Abstract

Due to increasing energy demand and depleting conventional energy sources, the utilization of environment friendly renewable energy sources (RES) is gaining importance in developing countries. Smart grid is the integration of existing power grid with renewable generation along with information and communication technology (ICT) which enables it to have various features like two-way power flow, self-healing, demand response, and energy management. Because of the intermittent nature of the renewable energy resources and the fluctuations in the load connected to the system, the forecasting of the load and RES play an important role in economical scheduling and optimal dispatch of the resources. In order to make the system reliable and self-healing, the continuous monitoring of the parameters at various locations in the network is mandate. Based on the parameters obtained at various points in a network, control will be done in centralized or decentralized manner and the appropriate decision will be sent from the control center to the intelligent electronic devices (IEDs). This chapter deals with the various challenges that are faced in large-scale integration of the RESs and the mitigation measures that are to be taken to overcome these challenges in smart grid.

*Corresponding author: klm.eee@psgtech.ac.in

M. Kathiresh A. Mahaboob Subahani and G.R. Kanagachidambaresan (eds.) Integration of Renewable Energy Sources with Smart Grid, (281–304) © 2021 Scrivener Publishing LLC

Keywords: Renewable energy sources (RES), intelligent electronic devices (IED), distributed generators (DGs), point of common coupling (PCC), short-term load forecasting (STLF), demand side response (DSR), electric vehicle (EV), phasor measurement unit (PMU)

13.1   Introduction

The large-scale integration of renewable energy sources (RESs) with the existing generation have several benefits, such as a reduction in greenhouse gases emissions, reliable power supply to critical loads, and optimal operation with reduced losses. But generation from RES can be variable and uncertain, in contrast to conventional generation due to various factors like variable wind speed, solar irradiation, power output fluctuations, disturbance due to converters harmonics, and low inertia of the DG. In order maintain the system stability with RES, sufficient reserve capacity that acts when the renewable resources drop below their anticipated production level must be in place to overcome load and renewable forecast errors or unexpected failures of generators [1]. Along with the reserve capacity, demand response (DR) and storage will also be required to cover the possible mismatches between demand and generation caused by weather-influenced power plants. By introducing intermittent generation and by avoiding fuel and variable costs of generation units that can be dispatched on demand, the overall marginal costs of the system tend to decrease [2]. Power converters are used for connecting the renewable to the existing grid, but as the level of converters increases, it is essential that all these new converters and existing ones operate together in an accepted manner. The main requirement for the generating units and the large-scale RESs is that they should stay connected to the grid at voltage drop in the connection point along with coordinated control and avoidance of adverse interactions in the grid. Voltage ride through capability of the converters at the PCC helps to improve the power system strength and the local voltage quality during faults in the system [3]. Protection systems of the power system which was designed for the unidirectional power flow will become obsolete because of the distributed generation, and as a consequence, detection of fault current becomes cumbersome and more sophisticated mitigation measures are needed [4]. Voltage quality of supply depends on the level of system strength, which, in turn, depends on the number, size, and location of synchronous units in the system. Many researchers focus on minimizing and eliminating harmonic and imbalance current contributions from the converters of the renewable energy to limit their potential negative influence on the voltage quality. A critical consideration is the level of voltage harmonics in the system [5]. Lowering the system fault level increases the level of voltage harmonics which leads to system imbalance and it is to be taken into consideration while designing a stable electric power grid. Significant imbalances could result in interconnection frequency deviations, transmission system violations, stability problems, etc., that ultimately could lead to widespread system blackouts. Variability and uncertainty in wind generation is managed mainly by using flexible resources in the power systems [6, 7]. Flexibility refers to the ability to change power output level or consumption according to the system needs when keeping the continuous balance of consumption and generation. A distributed energy management system (DEMS) combines elements of DR and distributed generation in a way that allows them to be dispatched optimally and intelligent control system plays an important role in voltage management of the grid, balancing and scheduling of renewables sources, and local grid load management [8, 9]. The development of the microgrid concept is presented as a solution to overcome some of the negative impacts of massive micro generation deployment [10]. Microgrid is low-voltage (LV) distribution systems with distributed energy resources (DERs) [microturbines, fuel cells, photovoltaic (PV), etc.], storage devices (batteries, flywheels), energy storage system, and flexible loads. Such systems can operate either connected or disconnected from the main grid. The operation of microsources in the network can provide benefits to the overall system performance, if managed and coordinated efficiently [11–13]. Microgrid central controller (MGCC) analyzes real-time data of loads, distributed generations, grid-connected switch information, and historical data such as loads and generating capacity. According to the renewable energy most absorption, constant power operation, and peak load shifting, MGCC selects the control strategies and coordinating controls distributed generations, energy storages and loads to maintain microgrid stability [14].

13.2 Microgrid

A microgrid is a collection of interconnected distributed generators (DGs), energy storage units (ESU), and loads and is connected to the main grid. It is considered as a single generator or controllable load by the grid. Microgrid can either work in islanded mode of operation or in grid-connected mode [11–13]. When there is disturbance occurring in the system such as voltage sags or faults, the microgrid isolates itself from the main grid along with the storage system and load connected to it. This is called as dynamic islanding and is done to isolate the load from disturbances. In grid connected mode, the microgrid is connected directly to the main grid. In this mode of operation, the main grid controls the voltage and frequency. During islanded mode of operation the microgrid controls the voltage and frequency.

13.2.1   Types of Microgrid

There are two types of microgrid, namely, AC microgrid and DC microgrid. The RESs such as fuel cells and PV cells produce DC, whereas wind energy system produces AC and is of variable frequency [10]. Hence, in case of AC microgrid, a DC/AC or AC/DC/AC power conversion stage is essential. This is totally eliminated in case of DC microgrid and has the advantage in terms of cost, efficiency, and size of the system. Also, another advantage of DC microgrid is that it does not require reactive power.

13.2.1.1   DC Microgrid

There are several loads like LED lights, computers, television, and several other consumer electronics which require DC power to operate. So, instead of wasting energy in converting DC to AC and AC to DC to AC and again to DC for operating such loads, the power output from a DC grid can be used directly. A schematic of a DC microgrid system is shown in Figure 13.1. One of the major advantages of DC microgrid is its capability of static storage integration. The storage element such as batteries and ultra-capacitors uses DC supply. The mechanical energy storage systems like flywheels which are coupled to PMSM are also connected to the distribution system through DC link. The addition of DC storage to a DC microgrid is comparatively simple than connecting a DC storage to an AC system as it requires extra hardware. The components of DC microgrid are i) distributed energy sources like wind turbine and fuel cells, ii) the distributed energy storage devices like batteries, and iii) critical and non-critical loads. The distributed energy storage devices compensates for the shortage or surplus power in the microgrid. A bidirectional power converter monitors the power flow between both the sides and the microgrid is connected to the macrogrid via the bidirectional power converter. On occurrence of

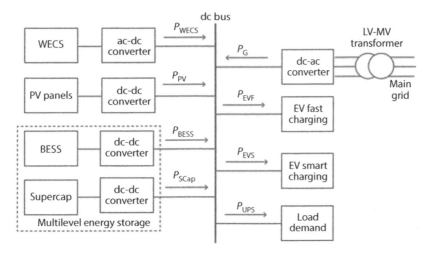

Figure 13.1 Schematic of a DC microgrid.

any fault in the utility grid, the microgrid will be disconnected from the grid by the power converter and the microgrid will operate in islanded mode. The point of common coupling (PCC) is the connection point to the main distribution utility. During the islanded operation mode, the DC-link voltage level is regulated by the battery and if there is any sudden change in power requirement the super capacitor responds to it as an auxiliary source or sag. The battery gets discharged so as to maintain the DC bus voltage, if the generated power is less than the load demand.

If the load demand is higher than the generated power, load shedding may be required. Based on priorities, the load shedding will be implemented. The loads with lowest priority are being tripped first and that with highest priority will be disconnected at the end.

13.2.1.2   AC Microgrid

The components of main AC microgrid are DERs, loads, master controller, protective devices, switches, control, and automation and communication systems. The schematic of an AC grid with multiple distributed energy system is shown in Figure 13.2. The voltage at PCC is determined by the main grid. In the grid connected mode, direct frequency and voltage control is not required instead the main grid will provide the reference frequency and voltage. In this mode, the major role of microgrid is to accommodate the load demand and the real and reactive power generated by the distributed generation units.

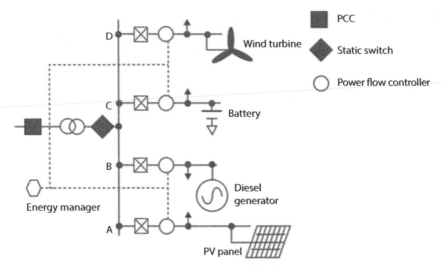

Figure 13.2 Schematic of AC microgrid with multiple distributed energy system.

13.2.1.3   Hybrid AC-DC Microgrid

The schematic of a hybrid AC-DC microgrid with multiple distributed energy system connected to the main grid is shown in Figure 13.3. In hybrid microgrid, both AC and DC buses are interconnected by using bidirectional AC-DC converter which is connected to the main grid at the PCC. It combines the advantages of both AC and DC microgrid.

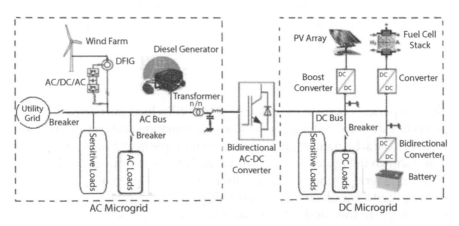

Figure 13.3 Schematic of hybrid AC-DC microgrid with multiple distributed energy system [15].

13.3 Large-Scale Integration of Renewables: Issues and Challenges

The generation is scheduled based on forecasted demand, then re-dispatched closer to delivery with updated information on forecasts, and the operating reserves acts to keep the balance of demand and generation throughout the grid. System operators follow the change in total demand, not the variation from a single generator or customer load. Wind energy forecasting can be used to predict wind energy variability in advance through a variety of methods based on numerical weather-prediction models and statistical approaches [17]. Both accuracy and uncertainty of short-term forecasts are important information for system operators, when allocating the reserves needed to manage the real-time operation [18]. With high wind penetration, it will be beneficial to allocate reserve requirements dynamically. If allocation is estimated once a day for the next day instead of using the same reserve requirement for all days, the low wind days will induce less requirements for the system, and thus, the reserve allocation can be increased only for days when wind variability and uncertainty are at highest [16–20]. But the forecast error in both demand and the RESs will create a larger dynamics in the optimum scheduling of the generation along with efficient dispatch issues.

The main issues while integrating the RESs to the electric grid are as follows:

- Impacts on balancing in different time scales: any increase needed in short-term reserves or ramping requirements [30], scheduling [29], and efficiency of conventional power plants;
- Impacts on grid reinforcement needs and grid stability [21, 22];
- Impacts on generation adequacy—how to meet the peak loads.

If the generation and demand in the system does not match, then it may lead to serious balancing issues and stability of the electrical grid will also be affected. The studies show a significant reduction of operational costs (fuel usage and costs) due to wind power even at higher penetration levels so the integration effort will not offset the emission savings of wind power [23, 24]. To capture the integration cost means capturing the difference of full credit for operating cost reduction compared with cost for system operation, with efficiency penalties due to increased variability and uncertainty. One way of capturing cost of variability is by comparing simulations with flat wind energy to varying wind energy [23]. However, the two simulated cases can also result in other cost differences than just the variability cost [24]. Challenging situations seen in system operation so far are from high wind-power generation during low-load situations, when wind power penetration levels exceed 50% [31]. Wind is usually last to be curtailed. However, when all other units are already at a minimum (and some shut down), system operators sometimes need to curtail wind power to control frequency which may lead to susceptibility to instability in the system [34–37]. The power system should sustain disturbances, such as the loss of largest power plant or line so that the frequency and voltage remain stable. Wind power is asynchronous generation that does not have the same inherent, physical support to the power system inertia as synchronous machines. For small-scale renewable integration the concept of microgrid has been introduced which is group of interconnected loads and DERs within clearly defined electrical boundaries that acts as a single controllable entity with respect to the grid. A microgrid can connect and disconnect from the grid to enable it to operate in both connected or island-mode [12]. There are various researches going on to represent various issues in microgrid for economic operation, scheduling, control, islanding, and monitoring [27, 28]. The intermittent nature of irradiance, primarily due to the variability of cloud cover, can cause significant fluctuations in the power delivered by large-scale grid-connected PV plants [32, 33]. In order to overcome these fluctuations, voltage and current phasors must be continuously transferred from PMU to a central data-processing system, known as a phasor data concentrator (PDC). These data are analyzed in the control center in decentralized or centralized manner to control and monitor the operation of the smart grid.

13.4   A Review on Short-Term Load Forecasting Methods

For deciding authority in the power sector, the decision-making process is more complex with different stages that have to be taken into account. It comprises the planning of facilities and an optimal day-to-day operation of the power generating station. These decisions are widely differ with respect to time-horizons and aspects of the system. For accomplishing such tasks, load forecasts are most important. This paper represents a complete survey of the short-term load forecasting (STLF). It also reviews the different topologies for STLF. This survey article ensures that it will be very much helpful to the researchers working in the field of STLF. The prevision of short-term loads, i.e., STLF plays a vital role to formulate the economic, reliable, and secure operating strategies for power network. Importunity of prevision is an important view in the design of any model for electricity planning, especially in today's restructuring the power system. The type of the demand laid on the kind of planning and precision that is required. It shows out that even accurate load forecasts could not guarantee returns. The market competition related to selling is affordable due to maximum volatility of electricity prices. In addition to that, secure operation of power system requires the study of its behavior under a various contingency conditions. Depending on the time zone, the load forecasting is divided in to following categories:

- Very short-term load forecasting
- Short-term load forecasting
- Mid-term load forecasting

The assessed forecasts in this time zone are important key notes for producing scheduling functions, power system security assessment and power system dispatcher. Due to reputation of the load forecasting, enormous topologies for STLF have been reported in the last few years. These methodologies can be abridged in to deterministic, stochastic, knowledge-based expert systems, and artificial neural networks (ANN) [47]. Use of above methodologies with fuzzy interface is also included in the literature. The objective of this review is to survey and differentiate electric load forecasting technology published till date. This not only covers the comparison of old literatures but also covers new categories of technologies that in recent researches. It also gives upgraded detailed description. STLF techniques are divided into four types. Those are as follows:

- Statistical Technique [50]
- Artificial Intelligent (AI) Technique [47, 48]
- Knowledge-Based Expert Systems [45, 46]
- Hybrid Techniques [52]

The deterministic approaches are conventional causal model of load and weather liable to change with respect to weather. This comprises curve fitting, data computation, and smoothing methods. The stochastic approaches model [49–51] the load performance in terms of stochastic process. Kalman filtering are autoregressive averages and time sequence approaches are coming under this category. Knowledge-based expert system is order of the day. The fruitful result of which has shown nowadays. These models are based on earlier load behavior. Fuzzy logic is also a good application for such systems. Application of artificial neural network commenced in early decades, since then, large amount of research work has gone in this area. The ANN-based forecasting techniques used in various technologies based on both supervised and unsupervised mode of learning. Feed forward network used in both single layer and multilayer in fully connected or non-fully connected technology. Recurrent networks also came into existing because of dynamic modeling of load with good outcomes.

13.4.1   Short-Term Load Forecasting Methods

A various methods including similar day approach, regression models, time sequences, neural networks, fuzzy logic, and many more methods are used for STLF. The selection of appropriate tools leads to more accurate in load forecasting.

13.4.1.1   Statistical Technique

This approach requires a clear mathematical model which interlinks between load and some input factors. Few classical models are manipulated for this load forecasting, such as:

- Multiple regression method
- Exponential smoothing
- Iterative reweighted least square
- Adoptive load forecasting
- Stochastic time series

13.4.1.1.1   Multiple Regression Method
The technique used in this method is weighted least squares estimation. With the help of this analysis the statistical relationship of total load, weather condition and the day type influences can be calculated. The regression coefficients are estimated using least squares estimation with defined amount of historical data.

13.4.1.1.2   Exponential Smoothing
These approaches are classical methods used for load forecasting. In exponential smoothing, firstly, load should be modeled with the observed previous data then the future load can be predicted.

13.4.1.1.3   Iterative Reweighted Least Square

This method used to find the model order and parameters. The operator used in this method will control only one variable at a time. The operator can also determine the optimal starting point. This approaches uses the autocorrelation function and the partial autocorrelation function to difference the past load data and to identify the suboptimal model of the load dynamics. The weighting function, tuning constants, and weighted sum of squared residuals create a tri-way decision variable in identifying an optimal model and the successive parameter estimation.

13.5   Overview on Control of Microgrid

13.5.1   Need for Microgrid Control

The RESs like wind turbine, solar system, biomass, and several other resources generate and feed power to the microgrid. So, the performance of these sources decides the performance of the microgrid. The performance of these renewable energy resources is dependent on several factors like sunlight and wind, and they vary with time. With these variations, the performance of the RESs get affected which, in turn, affects the microgrid performance which causes oscillations in voltage and frequency [42–44, 53, 54]. Since these uncertainties may cause grid failure, it has to be addressed to make the performance of the grid reliable and stable.

So, there should be certain control strategies to solve the issues listed below [55–60]:

(i)   Regulation of frequency and voltage under any operating condition;

(ii)   Reactive and active power control to achieve power sharing in both grid connected and islanded mode of operation;

(iii) Smooth transition from isolated mode to grid connected mode and vice versa;

(iv)   Providing uninterrupted power supply to the critical loads during islanded mode of operation;

(v)   In case of grid failure, the microgrid should have the capability to black start;

(vi)   Meeting of power quality demands during islanding mode.

The overall control of power system is done in two ways: i) centralized control and ii) De-centralized control.

13.5.2 Fully Centralized Control

In this type of control, communication is been established between the central controller and the microgrid components. With the help of this communication system, the central controller performs the necessary calculations and determines the control action for the units connected to it at a single point [61]. The overall layout of a centralized control of AC microgrid is shown in Figure 13.4.

13.5.3 Decentralized Control

In this type of control, a local controller controls each unit separately. The local controller receives information from the corresponding unit alone and it does not consider system wide variables. The components on the network can communicate with each other through the communication network in attaining the predominant objectives. In this type, the control is of peer-peer type like gossip-based algorithm [62].

Fully centralized control is highly complicated as it requires communication to a wider extent. Since there is strong coupling between various units in the system, fully decentralized control is also not possible. Therefore, it is desired that the controllers are distributed for various distributed resources to avoid single point failure and delay in communication. A trade-off between fully centralized and fully decentralized control

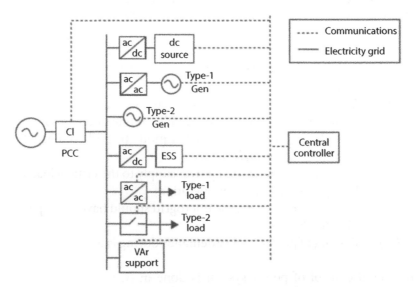

Figure 13.4 Schematic of centralized control of AC microgrid [61].

can be achieved by using hierarchical control. It involves three stages of control, namely, i) primary control ii) secondary control, and iii) tertiary control [63]. Layout of hierarchical control of an AC microgrid is shown in Figure 13.5. The load sharing between the converters in the microgrid is established by droop control which forms the primary control. The steady state error produced by the droop control is reduced in the secondary level of control. The last level, namely, tertiary level, exports or imports energy for the microgrid.

13.5.4   Hierarchical Control

13.5.4.1   Primary Control

Primary control is the first level of control in the hierarchical control and is also called as local control. The primary control is designed to satisfy the requirements listed below [64–67]:

- To stabilize the frequency and voltage. The microgrid may lose its stability with respect to voltage and frequency due

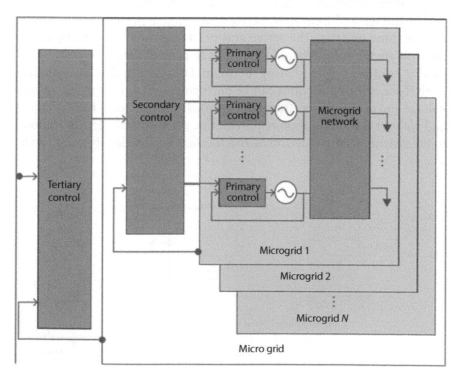

Figure 13.5 Hierarchical control of AC microgrid [63].

to mismatch between the consumed power and generated power after islanding.

- To share active and reactive power between the multiple DERs without any communication links.
- To mitigate circulating currents.

The primary control provides the voltage and current references for the control loops of DERs. These inner control loops are also referred as zero-level control. This control is implemented in either voltage control or PQ mode [68]. The main use of local controllers is to control DGs to operate in normal operation. The local controllers operate as a voltage controlled voltage source converter (VSC), where the droop controllers provide the voltage reference. In this voltage control mode, the nested voltage and current control loop are shown in Figure 13.6 [69–71]. This controller feeds the current signal as a feed forward term via a transfer function (e.g., virtual impedance). The proportional integral PI controllers are largely used for the design of the control loops in the local control level. It may be used alone or with feed forward compensation in closed loop systems to enhance the performance of the system. Apart from control of voltage and frequency, active and reactive power control should also be done by DGs. The active and reactive power controls by droop controls are the most common methods to control these powers. Similar to this control, power sharing between DGs could also be included. Both these control functions can be performed by using reactive power–voltage and active power–frequency droops based on local measurement and without communication requirements. For the controllers to respond to the systems' inherent dynamics and transients, a detailed dynamic model

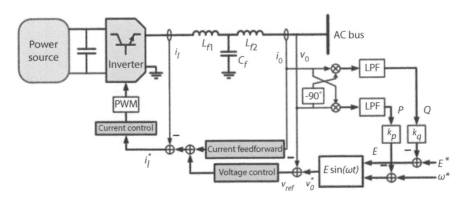

Figure 13.6 Local control loops in a typical voltage-controlled VSC-based DG [71].

of the DGs, including the resistive, reactive, and capacitive local load and the distribution system should be adapted.

13.5.4.2   Secondary Control

The secondary control helps in maintaining the stability of the grid operational parameters. After every change in load or generation, the secondary control ensures that the voltage and frequency deviations are regulated toward zero. Apart from this, the control can be also used for performing microgrid synchronization to the main grid before performing the interconnection, transiting from islanded to grid-connected mode. Compared with the primary control, the initial time of secondary control is higher because of the availability of primary resources and also the capacity of the energy storage. Compared to the primary control, this control takes minutes for (i) decoupling secondary control from primary control, (ii) reducing the bandwidth for communication by using microgrid variables, and (iii) allowing enough time in performing calculations. The amplitude and frequency levels in the microgrid are sensed and compared with the reference amplitude and frequency. This produces the error which is then processed through compensators and sent to all the units to restore the output voltage. A central controller is required to ensure the seamless operation of the power system during major disturbances such as transition from grid-connected mode to islanded mode [69–71].

13.5.4.3   Tertiary Control

Tertiary control ensures economically optimal operation, based on energy cost and electricity market. In order to optimize the microgrid operation within the utility grid, this controller enhances exchange of information or data with the distribution system operator. In grid connected mode of operation, the power flow is controlled by adjusting the frequency and amplitude of the voltage at the PCC [72–73]. Before synchronization, the process starts from the microgrid references with the frequency and amplitude of the mains grid. After the synchronization, these signals are given by the tertiary control. Tertiary control is considered as a part of the host grid and not the microgrid itself. This control is also responsible for coordinating the operation of multiple microgrids interacting with one another in the system. For example, the overall reactive power management of a grid could be accomplished by properly coordinating all the microgrids connected to it, through a tertiary control approach.

This control level typically takes several minutes, in providing signals to secondary level controls present at microgrids and other sub-systems that form the full grid.

13.6   Measures to Support Large-Scale Renewable Integration

Continuously increasing percentage of renewable source integrated into a power system presents additional challenges to system operators, especially in small interconnected power systems [34]. There are several methods that could be used to reduce the forecast errors that cause main problems for large-scale wind integration which are listed below:

- Aggregated thermal storages for balancing of power generation forecast errors [78].
- Pumped hydro energy storage for balancing of power generation forecast errors [74].
- Demand side management for providing balancing power [24–27].
- EV charging as alternative storage for Renewable Energy [75–77].

The probability distribution function to model the prediction errors of wind speed and solar radiation which establishes a two-stage stochastic optimization scheduling model with the objective of minimizing expected operating costs [28]. Therefore, DR is introduced to smooth the fluctuations in renewable energy, providing a new and important way to resolve the large-scale integration and efficient operation of RESs [79, 80]. An energy management system (EMS) is proposed to optimize the operation of REG. Through the integration of demand-side management (DSM) and an active management scheme (AMS), an EMS can make good use of renewable energy and reduce customer energy consumption costs while increasing the economic and environmental benefits [24–27, 80]. A bi-layer coordinated generation and reserve schedule method is proposed to for a grid-connected MG in a multi-time-scale framework to reduce the impact of uncertainties on power balance, operating costs, and system reliability. A relaxed reserve configuration model with bidirectional boundaries constraints is proposed to cope with the uncertainties in the optimal scheduling problem [81]. Own-price and cross-price elasticity models are proposed to integrate the short-term DR into the unit commitment (UC) optimization model for DR and renewable energy joint scheduling [41, 84]. DR model based on game theory is established and applied to a multiobjective dynamic economic dispatch [38]. Using the idea of rolling scheduling, it establishes a joint decision model for renewable energy and DR based on day-ahead and real-time time scales, in which the day-ahead optimization results are applied to real-time scheduling decisions [24, 28, 39, 40]. ESS provides a promising alternative in helping utilities to facilitate the integration of renewable energy and improve the grid reliability and efficiency [82]. They can operate in any one of the below mentioned perspective:

- ESS operation with unit UC to provide reserve, shave peak load, minimize the total generation cost, and avoid extra start-up and shut-down operating cost [83, 84];
- ESS planning and operation with optimal power flow (OPF) to alleviate the feeder flow capacity limitation and reduce line loss [85];
- Combining ESS participation with UC and OPF together

To tackle the problems of renewable integration, preventive control of wind power is required on the basis of the original security and stability control system and is complement with event-based control function.

13.6.1   Basic Idea of Preventive Control

The preventive control of wind power, unlike thermal power or hydropower, is hard to implement as the output of the wind unit cannot strictly follow the generation schedule set by grid dispatcher all the time. There are two modes for wind power preventive control [86].

13.6.1.1   Maximum Output Control Mode

Under condition that grid is secure and stable, calculate the maximum generation output for each wind farms according to grid's wind power acceptance capability. When wind generator's output is below its upper limit, it is in the free generation state, while ramp-up rate must meet certain requirements. When the generator's output exceeds the upper limit, it can take up system resources of other wind farms according to other fields' availability, to achieve the goal of maximum generation and fully utilization of wind resources among wind farms.

13.6.1.2   Output Following Mode

According to prediction of wind power of each wind farm, schedule generation plan for each wind farm after security evaluation, and each wind farm must follow the plan by making real-time active power output adjustment. The two control modes are suitable for preventive control of several wind farms in a region, as well as coordinated preventive control of wind farm groups in different regions.

References

1. González-Aparicio, I. and Zucker, A., 'Impact of wind power uncertainty forecasting on the market integration of wind energy in Spain'. *Appl. Energy*, 159, 334–349, 2015.
2. *Energy Prices and Costs in Europe*, European Commission, Brussels, 2014.
3. Urdal, H., Ierna, R., Zhu, J., Ivanov, C., Dahresobh, A., Rostom, D., System strength considerations in a converter dominated power system. *IET Renewable Power Gener.*, 9, 1, 10–17, 2015.
4. Salomonsson, D., Soder, L., Sannino, A., Protection of Low-Voltage DC Microgrids. *IEEE Trans. Power Delivery*, 24, 3, 1045–1053, 2009.
5. Akbarali, M.S., Subramanium, S.K., Natarajan, K., Real and reactive power control of SEIG systems for supplying isolated DC loads. *J. Inst. Eng. India Ser. B*, 99, 6, 587–595, 2018.
6. Berry, M., Cornforth, D.J., Platt, G., An introduction to multiobjective optimisation methods for decentralized power planning. *Power Energy Soc. General Meet*, 2009, https://doi.org/10.1109/PES.2009.5275716.
7. Akbarali, M.S., Subramanium, S.K., Natarajan, K., Modeling, analysis, and control of wind-driven induction generators supplying DC loads under various operating conditions. *Wind Eng.*, 1–14, 2020, https://doi.org/10.1177/0309524X20925398.
8. Melsas, R., Rosin, A., Drovtar, I., Wind park cost efficiency increase through direct cooperation with demand side response provider. *2016 57th International Scientific Conference on Power and Electrical Engineering of Riga Technical University (RTUCON)*, Riga, pp. 1–5, 2016.
9. Pudjianto, D., Ramsay, C., Strbac, G., Virtual power plant and system integration of distributed energy resources. *IET Renew. Power Gener.*, 1, 10–16, 2007.
10. Lopes, J.A.P., Madureira, A.G., Moreira, C., "A View of Microgrids", Advances in Energy Systems: The Large-scale Renewable Energy Integration Challenge, John Wiley & Sons Ltd, USA, 2019.
11. Ackermann, T., *Wind Power in Power Systems*, Second Edition, 1049 pp, John Wiley & Sons Ltd, The Atrium, Southern Gate, Chichester, West Sussex PO19 8SQ, England, 2012.

12. Lasseter, R.H., MicroGrids, in: *2002 IEEE Power Engineering Society Winter Meeting. Conference Proceedings (Cat. No. 02CH37309)*, vol. 1, pp. 305–308.
13. Hatziargyriou, N., *Microgrids Architectures and Control*, p. 4, John Wiley and Sons Ltd, USA, 2014.
14. Tian, B., Lei, J., Guo, X., Ang, P.Y., Yuan, H., Xu, Z., Zhou, S., He, T., Design of MGCC for User-side Microgrid. *International Conference on Power Electronics Systems and Applications*, Hong Kong, China, 2015.
15. Kaushik, R.A. and Pindoriya, N.M., A Hybrid AC-DC Microgrid: Opportunities & Key Issues in Implementation. *International Conference on Green Computing Communication and Electrical Engineering (ICGCCEE)*, Coimbatore, 2014.
16. de Almeida, R.G., Castronuovo, E.D., Lopes, J.A.P., Optimum generation control in wind parks when carrying out system operator requests. *IEEE Trans. Power Syst.*, 21, 718–725, 2006.
17. Giebel, G., Brownsword, R., Kariniotakis, G. *et al.*, *The state of the art in short-term prediction of wind power: a literature overview. Deliverable D-1.2 of Anemos.plus*, ANEMOS, Wilhelmshaven, Germany, 2011, Available at: http://www.prediktor.dk/publ/GGiebelEtAl-StateOfTheArtInShortTermPrediction_ANEMOSplus_2011.pdf. (Accessed May 14, 2012).
18. Ela, E. *et al.*, Evolution of operating reserve determination in wind power integration studies, in: *Proc. IEEE Power Energy Soc. (PES) Gen. Meeting*, pp. 1–8, Minneapolis, MN, USA, Jul. 2010.
19. He, M., Yang, L., Zhang, J., Vittal, V., A spatio-temporal analysis approach for short-term forecast of wind farm generation. *IEEE Trans. Power Syst.*, 29, 4, 1611–1622, Jan. 2014.
20. Durrwachter, H.L. and Looney, S.K., Integration of wind generation into the ERCOT market. *IEEE Trans. Sustain. Energy*, 3, 4, 862–867, Oct. 2012.
21. Cardell, J. and Anderson, C.L., A flexible dispatch margin for wind integration. *IEEE Trans. Power Syst.*, 30, 3, 1–10, May 2015.
22. Ortega-Vazquez, M.A. and Kirschen, D.S., Estimating the spinning reserve requirements in systems with significant wind power generation penetration. *IEEE Trans. Power Syst.*, 24, 1, 114–124, Feb. 2009.
23. De Jonghe, C., Hobbs, B.F., Belmans, R., Optimal generation mix with short-term demand response and wind penetration. *IEEE Trans. Power Syst.*, 27, 2, 830839, May 2012.
24. Sahin, C., Shahidehpour, M., Erkmen, I., Allocation of hourly reserve versus demand response for security-constrained scheduling of stochastic wind energy. *IEEE Trans. Sustain. Energy*, 4, 1, 219228, Jan. 2013.
25. Meibom, P., Weber, C., Barth, R. *et al.*, Operational costs induced by fluctuating wind power production in Germany and Scandinavia. *Renew. Energy Gener.*, 3, 75–83, 2009.
26. Eftekharnejad, S., Vittal, V., Heydt, G., Keel, B., Loehr, J., Small signal stability assessment of power systems with increased penetration of photovoltaic

generation: A case study. *IEEE Trans. Sustain. Energy*, 4, 4, 960–967, October 2013.

27. Jiang, Q., Xue, M., Geng, G., Energy Management of Microgrid in Grid-Connected and Stand-Alone Modes. *IEEE Trans. Power Syst.*, 28, 3380–3389, 2013.

28. Jeylani, M.A., Mahaboob Subahani, A., Kanakaraj, J., Rameshkumar, K., Analysis of Solar PV Application based on Bidirectional Inverter. *Proc. META Web of Conferences*, September 2018, vol. 225, pp. 1–6, 2018.

29. Li, P., Guan, X., Wu, J., Zhou, X., 'Modeling dynamic spatial correlations of geographically distributed wind farms and constructing ellipsoidal uncertainty sets for optimization-based generation scheduling'. *IEEE Trans. Sustain. Energy*, 6, 4, 1594–1605, 2015.

30. Lawrence Berkeley National Laboratory, Mass market demand response and variable generation integration issues: a scoping study. LBNL report LBNL-5063E, Environmental Energy Technologies Division, USA, October 2011.

31. Hoff, T.E. and Perez, R., Quantifying PV power output variability. *Sol. Energy*, 84, 1782–1793, 2010.

32. Marcos, J., de la Parra, Í., García, M., Marroyo, L., Simulating the variability of dispersed large PV plants. *Prog. Photovoltaics: Res. Appl.*, 24, 680–691, 2016.

33. Haaren, R., van Haaren, R., Morjaria, M., Fthenakis, V., Empirical assessment of short-term variability from utility-scale solar PV plants. *Prog. Photovoltaics: Res. Appl.*, 22, 548–559, 2014.

34. Jupe, S.C.E., Michiorri, P.C., Taylor, A., Coordinated output control of multiple distributed generation schemes. *Renew. Power Gener.*, 4, 283–297, 2010.

35. Burke, D.J. and O'Malley, M.J., Factors influencing wind energy curtailment. *IEEE Trans. Sustain. Energy*, 2, 2, 185–193, 2011.

36. Fink, S., Mudd, C., Porter, K., Morgenstern, B., Wind energy curtailment case studies, NREL Subcontract Report, NREL/SR-550, National Renewable Energy Laboratory, USA, vol. 46716, pp. 1–47, 2009.

37. Yasuda, Y., Bird, L., Carlini, E.M. *et al.*, International comparison of wind and solar curtailment ratio, in: *International Workshop on Large-Scale Integration of Wind Power into Power Systems as well as on Transmission Networks for Offshore Wind Farms*, October 2015, Energynautics GmbH, Brussels, Belgium, pp. 1–6.

38. Nwulu, N. I., and Xia, X., Multi-objective dynamic economic emission dispatch of electric power generation integrated with game theory based demand response programs. *Energy Convers. Manage.*, 89, 963974, Jan. 2015.

39. Galvan, E., Alcaraz, G.G., Cabrera, N.G., Two-phase short term scheduling approach with intermittent renewable energy resources and demand response. *IEEE Lat. Am. Trans.*, 13, 1, 181187, Jan. 2015.

40. Zakariazadeh, A., Jadid, S., Siano, P., Stochastic operational scheduling of smart distribution system considering wind generation and demand

response programs. *Int. J. Electr. Power Energy Syst.*, 63, 218225, Dec. 2014.

41. Wang, Y., Xia, Q., Kang, C., Unit commitment with volatile node injections by using interval optimization. *IEEE Trans. Power Syst.*, 26, 3, 17051713, Aug. 2011.

42. Hamzeh, M., Ghafouri, M., Karimi, H., Sheshyekani, K., Guerrero, J.M., Power Oscillations Damping in DC Microgrids. *IEEE Trans. Energy Convers.*, 31, 3, 970–980, 2016.

43. Magdy, G., Mohamed, E.A., Shabib, G., Elbaset, A.A., Mitani, Y., Microgrid Dynamic Security Considering High Penetration of Renewable Energy. *Prot. Control Mod. Power Syst.*, 3, 1, 23, 2018.

44. Jung, J. and Broadwater, R.P., Current status and future advances for wind speed and power forecasting. *Renewable Sustainable Energy Rev.*, 31, 762–777, 2014.

45. Ren, Y., Suganthan, P.N., Srikanth, N., Ensemble methods for wind and solar power forecasting—A state-of-the-art review. *Renewable Sustainable Energy Rev.*, 50, 82–91, 2015.

46. Foley, A.M., Leahy, P.G., Marvuglia, A., McKeogh, E.J., Current methods and advances in forecasting of wind power generation. *Renewable Energy*, 37, 1–8, 2012.

47. Cadenas, E. and Rivera, W., Short term wind speed forecasting in La Venta, Oaxaca, México, using artificial neural networks. *Renewable Energy*, 34, 274–278, 2009.

48. Li, G., Shi, J., Zhou, J., Bayesian adaptive combination of short-term wind speed forecasts from neural network models. *Renewable Energy*, 36, 352–359, 2011.

49. Bivona, S., Bonanno, G., Burlon, R., Gurrera, D., Leone, C., Stochastic models for wind speed forecasting. *Energy Convers. Manage.*, 52, 1157–1165, 2011.

50. Zhang, Y., Wang, J., Wang, X., Review on probabilistic forecasting of wind power generation. *Renewable Sustainable Energy Rev.*, 32, 255–270, 2014.

51. Liu, H., Tian, H., Li, Y., Comparison of two new ARIMA-ANN and ARIMA Kalman hybrid methods for wind speed prediction. *Appl. Energy*, 98, 415–424, 2012.

52. Liu, H., Chen, D., Tian, H., Li, Y., A hybrid model for wind speed prediction using empirical mode decomposition and artificial neural networks. *Renewable Energy*, 48, 545–556, 2012.

53. Akbarali, M.S., Subramanium, S.K., Natarajan, K., Modeling, analysis, and control of wind-driven induction generators supplying DC loads under various operating conditions. *Wind Eng.*, SAGE Publications, 1–4, 2020.

54. Vandoorn, T.L., Renders, B., Degroote, L., Meersman, B., Vandevelde, L., Voltage Control in Islanded Microgrids by Means of a Linear-Quadratic Regulator, in: *Proc. IEEE Benelux Young Researchers Symposium in Electrical Power Engineering (YRS'10)*, Belgium: Leuven, 2010.

55. Laria, A., Curea, O., Jiminez, J., Camblong, H., "Survey on microgrids: unplanned islanding and related inverter control techniques". *Renewable Energy*, 36, 2052–61, 2011.

56. Blaabjerg, F., Teodorescu, R., Liserre, M., Timbus, A.V., Overview of Control and Grid Synchronization for Distributed Power Generation Systems. *IEEE Trans. Ind. Electron.*, 53, 5, 1398, 1409, Oct. 2006.

57. Guerrero, J.M., Chandorkar, M., Lee, T., Loh, P.C., Advanced Control Architectures for Intelligent Microgrids—Part I: Decentralized and Hierarchical Control. *IEEE Trans. Ind. Electron.*, 60, 4, 1254, 1262, April 2013.

58. Miveh, M.R., Rahmat, M.F., Mustafa, M.W., A new per-phase control scheme for three-phase four-leg grid-connected inverters. *Electron. World*, 120, 1939, 30–36, 2014.

59. Lasseter, R.H., Microgrids and Distributed Generation. *Intell. Autom. Soft Comput.*, 16, 2, 225–234, 2010.

60. Lopes, J.A.P., Moreira, C.L., Madureira, A.G., Defining control strategies for Micro Grid-islanded operation. *IEEE Trans. Power Syst.*, 21, 2, 916–924, 2006.

61. Meng, L., Savaghebi, M., Andrade, F., Vasquez, J.C., Guerrero, J.M., Graells, Microgrid Central Controller Development and Hierarchical Control Implementation in the Intelligent Microgrid Lab of Aalborg University, in: *Applied Power Electronics Conference and Exposition (APEC)*, 2015.

62. Zamora, R. and Srivastava, A.K., Controls for Microgrids with Storage: Review, Challenges, and Research Needs. *Renewable Sustainable Energy Rev.*, 14, 7, 2009–2018, 2010.

63. Olivares, D.E., Mehrizi-Sani, A., Etemadi, A.H., Cañizares, C.A., Iravani, R., Kazerani, M., Hajimiragha, A.H., Gomis-Bellmunt, O., Saeedifard, M., Palma-Behnke, R. *et al.*, Trends in Microgrid Control. *IEEE Trans. Smart Grid*, 5, 4, 1905–1919, 2014.

64. Guerrero, J.M., Vásquez, J.C., Matas, J., Castilla, M., Vicuña, L.G.D., Castilla, M., Hierarchical control of droop-controlled AC and DC microgrids—A general approach toward standardization. *IEEE Trans. Ind. Electron.*, 58, 158–172, Jan. 2011.

65. Mehrizi-Sani, A. and Iravani, R., Potential-function based control of a microgrid in islanded and grid-connected models. *IEEE Trans. Power Syst.*, 25, 1883–1891, Nov. 2010.

66. Brabandere, K.D., Vanthournout, K., Driesen, J., Deconinck, G., Belmans, R., Control of microgrids, in: *Proc. IEEE Power & Energy Society General Meeting*, pp. 1–7, 2007.

67. Nikkhajoei, H. and Lasseter, R.H., Distributed generation interface to the CERTS microgrid. *IEEE Trans. Power Del.*, 24, 1598–1608, Jul. 2009.

68. Karimi, H., Nikkhajoei, H., Iravani, M.R., Control of an electronically-coupled distributed resource unitsubsequent to an islanding event. *IEEE Trans. Power Del.*, 23, 493–501, Jan. 2008.

69. Lopes, J.A.P., Moreira, C.L., Madureira, A.G., Defining control strategies for microgrids islanded operation. *IEEE Trans. Power Syst.*, 21, 916–924, May 2006.

70. IEEE-PES Task Force on Microgrid Control Trends in microgrid control. *IEEE Trans. Smart Grid*, 5, 4, 1905–1919, 2014.

71. Bidram, A. and Davoudi, A., Hierarchical structure of microgrids control system. *IEEE Trans. Smart Grid*, 3, 4, 1963–1976, 2012.

72. Bevrani, H., Watanabe, M., Mitani, Y., Microgrid controls, in: *Standard Handbook for Electrical Engineers*, 16th edn, H. Wayne Beaty (Ed.), pp. 160–176, McGraw-Hill, Pennsylvania Plaza, USA, 2012, Section 16.9.

73. Logenthiran, T., Srinivasan, D., Khambadkone, A.M., Aung, H.N., Multiagent System for Real-Time Operation of a Microgrid in Real-Time Digital Simulator. *IEEE Trans. Smart Grid*, 3, 2, 925–933, June 2012.

74. Huertas-Hernando, D., Farahmand, H. *et al.*, Hydropower Flexibility for Power Systems with Variable Renewable Energy Sources: An IEA Task 25 Collaboration, in: *Advances in Energy Systems: The Large-scale Renewable Energy Integration Challenge*, John Wiley & Sons Ltd, Brooklyn, NY, 2019.

75. Galus, M.D., Waraich, R.A., Noembrini, F. *et al.*, Integrating power system, transportation systems and vehicle technology for electric mobility impact assessment and efficient control. *IEEE Trans. Smart Grid*, 3, 2, 934–949, 2012.

76. Weiller, C., Plug-in hybrid electric vehicle impacts on hourly electricity demand in the United States. *Energy Policy*, 39, 3766–3778, 2011.

77. Kempton, W. and Tomic, J., Vehicle-to-grid power implementation: from stabilizing the grid to supporting large-scale renewable energy. *J. Power Sources*, 144, 280–294, 2005.

78. Moreno-Munoz, A. *et al.*, *Large Scale Grid Integration of Renewable Energy Sources*, The Institution of Engineering and Technology, London, 2017.

79. Aghaei, J. and Alizadeh, M.-I., Demand response in smart electricity grids equipped with renewable energy sources: A review. *Renewable Sustainable Energy Rev.*, 18, 6472, Feb. 2013.

80. Zeng, B., Zhang, J., Yang, X., Wang, J., Dong, J., Zhang, Y., Integrated planning for transition to low-carbon distribution system with renewable energy generation and demand response. *IEEE Trans. Power Syst.*, 29, 3, 11531165, May 2014.

81. Lei, X., Huang, T., Yang, Y., Fang, Y., Wang, P., A Bi-layer Multi-Time Coordination Method for Optimal Generation and Reserve Schedule and Dispatch of a Grid-Connected Microgrid. *IEEE Access*, 7, 44010–44020, 2019.

82. Sumanth, Y., Jain, A., Das, P., Bhakar, R., Mathur, J., Kushwaha, P., Operational Strategy of Energy Storage to Address Day-Ahead Scheduling Errors in High RE Scenario. *8th IEEE India International Conference on Power Electronics (IICPE)*, Jaipur, 2018.

83. Hu, Z., Zhang, S., Zhang, F. *et al.*, SCUC with battery energy storage system for peak-load shaving and reserve support. *Proc. IEEE PES General Meeting*, pp. 1–5, 2013.
84. Pozo, D., Contreras, J., Sauma, E.E., Unit commitment with ideal and generic energy storage units. *IEEE Trans. Power Syst.*, 29, 6, 2974–2984, 2014.
85. Gayme, D. and Topcu, U., Optimal power flow with large-scale storage integration. *IEEE Trans. Power Syst.*, 28, 2, 709–717, 2013.
86. Alburguetti, L.M., Grilo, A.P., Ramos, R.A., Preventive Control for Voltage Stability Enhancement Using Reactive Power from Wind Power Plants. *IEEE Power & Energy Society General Meeting (PESGM)*, 2019.

14

Mitigation Measures for Power Quality Issues in Renewable Energy Integration and Impact of IoT in Grid Control

Hepsiba D.[1], L.D. Vijay Anand[2], Granty Regina Elwin J.[3],
J.B. Shajilin[4] and D. Ruth Anita Shirley[5*]

[1]*Department of Biomedical Engineering, Karunya Institute of
Technology and Sciences, Coimbatore, India*
[2]*Department of Robotics Engineering, Karunya Institute of
Technology and Sciences, Coimbatore, India*
[3]*Department of Information Technology, Sri Krishna College of
Engineering and Technology, Coimbatore, India*
[4]*Francis Xavier Engineering College, Tirunelveli, India*
[5]*Department of Electronics and Communication Engineering,
Sri Krishna College of Engineering and Technology, Coimbatore, India*

Abstract

Power quality issues are a major concern in renewable energy generation. This chapter discusses the diverse power quality issues and its impact in renewable energy generation. The power quality issues like power frequency variation, voltage fluctuation, waveform distortion, voltage imbalance, short-duration voltage variation, long-duration voltage variation, and transients are analyzed. In particular, the various power quality issues that are involved in wind and solar energy generation are discussed in detail. The impact of these issues and also the severity of the issue is examined in this paper. There are two methodologies involved in mitigating power quality issues in renewable energy—from the utility side or from the customer side. The various means of mitigating these problems devices like UPQC, DVR, D-STATCOM, and UPS. The impact of Internet of Things in controlling the grid system during the time of power issues is also examined. Further, the various standards that are used in identifying and solving power quality issues are also elaborated in this chapter. It is observed that in the

Corresponding author: ruthshirley06@gmail.com

M. Kathiresh A. Mahaboob Subahani and G.R. Kanagachidambaresan (eds.) Integration of Renewable Energy Sources with Smart Grid, (305–326) © 2021 Scrivener Publishing LLC

DC systems, there are two major advantages, namely, high reliability and simple arrangement despite the power quality issues faced.

Keywords: Renewable energy, power quality issues, mitigation, DC system, Internet of Things

14.1   Introduction

A man-invented luxury that is enjoyed by all, to the point of necessity, is electricity. There is high demand for energy and the sources of energy have also increased over the years to meet the requirement. As power industries move along the track of development with Smart Industrial Revolution 4.0, the number of non-linear and linear loads that are used has increased dramatically. AC grid proves to be the source of current harmonics as well as reactive power for the solid-state switching converters to draw reactive power from. This leads to a number of interference and disturbance in everyday activities. In fact, continuous power supply has become a crucial factor in communication systems, computer power supplies, power systems, equipment, energy, services like furnaces, high-voltage systems, electrical power generations, renewable energy systems, etc. Research and experimentation is being carried out to find an optimal solution for power quality problems such as power system grid intrusion, excess neutral current, load balancing, system balancing, and harmonics.

Centralized facilities use large-scale generation of electricity by means of wind farms, hydroelectric dams, nuclear power plants, fossil-fuel power plants, etc. However, in recent years, many shortcomings in power generation have risen such as shortage of fossil fuel has risen, loss of power due to long transmission lines, as well as increased emission of electrical power. "Power quality" defines the capacity of a system to produce a sinusoidal wave which is noise-free and extremely stable (considering frequency and voltage) of a perfect power supply. But, due to practical considerations such as electromagnetic interference and external disturbances due to load, it is not possible to build an ideal power supply. In view of this, a survey was carried out during the 1990s [1], characterizing the time period of various disturbances and how it affects power supply as shown in Figure 14.1. Voltage stability is a crucial issue in power quality that needs regulation. Similarly, voltage regulation is affected by the excitation system's time constant and balance of reactive power.

Harmonic values have a negative impact on the voltage stability, according to a survey made on wind farms which had 12 turbines [2].

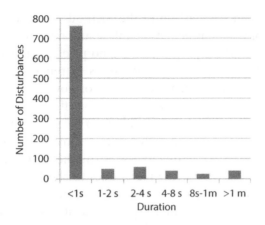

Figure 14.1 Distribution of power quality issues against time.

An examination of the duration of PQ disturbances showed that the number of interferences which exceeded one second were quite less when compared to the ones that lasted less than a second [3]. IEC and IEEE have also introduced standards to define power quality in 2011 [4]. Using distributed energy, it is possible to produce renewable energy which is purer and more efficient in terms of power transmission and generation [5]. A smart grid system, implemented with IoT-based cloud computing will also enable monitoring of power generation from a remote location. Spinning Reserve (SR) is the allocated capacity of energy that is reserved in a typical electric power system to employ custom power devices (CPDs) and to maintain quality of power [6]. Power quality issues occur in a number of places like renewable generation collector systems, residential power systems, and data centers that are used in DC distribution systems [7]. However, there are a number of issues that occur in affecting the power quality of which harmonics is the most troublesome one.

In [8], a fuzzy controlled UPQC, in DFIG-based grid connection is proposed to address the harmonics issue. Similarly in [9], FACTS is used to improve the power quality using wind energy systems. To improve power factor and reactive power consumption, Static Synchronous Compensator (STATCOM) is used. The implementation of Uninterruptible Power Supply is also a good option to reduce power quality issues. A UPS is used in many devices to protect electronic devices such as telecommunication devices and computers and can also be used as a standby system for many devices [10].

This book chapter will aim at establishing a study of the various ways that are used in mitigation of power quality issues in renewable energy sources. A detailed comparison of the methods are also studied and the best methodology is identified. A reliable solution for issues such as flicker, swell, and sag is UPQC [11]. There are no prior reference papers which examine the effect of power quality issues in renewable energy supply. Further, we have also studied the severity of power quality and how the proposed mitigation methods will prove to be effective. It will also be useful in understanding which issue is likely to have a major implication on power quality in the near future. This chapter aims at providing a complete analysis on the major power quality issues and the various methods of mitigating these issues. Monitoring techniques to check on power quality as well optimal solutions are also studied in an elaborate manner.

This book chapter focuses on the power quality issues and provides mitigation measures for the same. In the next session, power quality issues like power frequency variation, voltage fluctuation, waveform distortion, voltage imbalance, long and short-duration voltage variation, and transients are analyzed [12]. Section 14.3 gives an outline of the mitigating devices used such as D-STATCOM [14], UPQC [13], UPS, DVR, and TVSS. The outcomes are discussed in Section 14.4. Section 14.5 gives a conclusion for the power quality issues and Section 14.6 gives a brief view of future scope of power quality mitigation in renewable energy sources [15].

14.2   Impact of Power Quality Issues

Maintaining the frequency and magnitude of near sinusoidal rated current and voltage is known as "power quality". If there is any external disturbance that results in interrupting power quality, it will result in disrupting the efficiency of the system. In general, power quality control is said to be equivalent to voltage control [16]. This is primarily because it is easier to control the voltage when compared with current. Quality of power is defined using specifications such as harmonic content, transient currents and voltages, variation in voltage magnitude, and continuity of service. Wastage of economy and power is more when the quality of power is poor, which will affect the financial expertise of consumers and suppliers [17]. Unstable frequency and voltage will disrupt the quality of power that passes through a typical transmission line. Figure 14.2 represents power quality problems and how it can be evaluated by means of a flow diagram.

Power quality issues can be categorized into harmonic distortion, transients, flicker, voltage interruptions, and voltage unbalance [18]. The steps involved in evaluating the power quality issue are as follows:

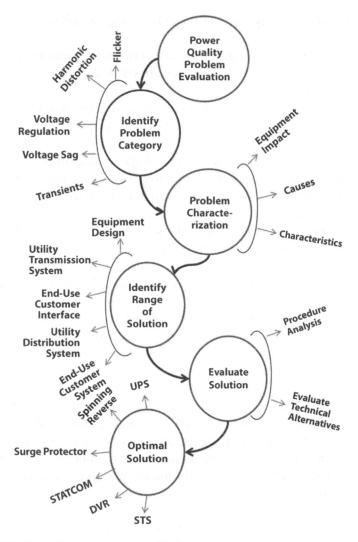

Figure 14.2 Flow diagram of power quality issues.

- First, the category of power quality issue is identified.
- The next step involves characterization of the problem by collecting data and measuring it to determine the equipment impact, characteristics, and causes of the problem.
- A number of solutions are then identified, suitable to fix the issue, and evaluation is also done accordingly.
- The last step is the evaluation of the final solution in order to obtain optimal economic results.

The major power quality issues [19] can be defined as follows:

- When there is voltage fluctuation between the range of 90%–110%, it is known as flicker, which will destroy the load side equipment.
- When the power system waveform is superimposed with a high frequency signal, it is known as "noise" which may also lead to loss of data.
- A decrease of load current or voltage or total interruption that for less than a second is referred to as "long time voltage interruption". This will lead to malfunctioning of data processing devices.
- Changes in voltage amplitude is referred to as voltage fluctuation. This occurs majorly because of sudden surges that happen in steady-state current, voltage or even both.
- A decrease in supply voltage magnitude is known as voltage sag.
- An increase in voltage, above the pre-defined level of tolerance, momentarily is called voltage swell. Voltage swell occurs for a period less than a few seconds and has a duration greater than a cycle.

A deviation from the ideal sine waveform is known as waveform distortion. Based on the deviation that occurs in spectral content, the waveform is characterized. In general, there are four types of waveform distortions, namely, noise, notching, interharmonics, and harmonics. Any variation in power frequency from its normal fundamental frequency will result in a form of distortion. The impact of nonlinear loads resulting in a waveform distortion of the currents and/or voltages is known as harmonics [20]. A spectral component is said to be harmonic under the following conditions if

$$f = nf_{Fund} \tag{14.1}$$

It will be a subharmonic component if

$$f > 0 \ and \ f < nf_{Fund} \tag{14.2}$$

It will be known as an interharmonic component if

$$f = nf_{Fund} \tag{14.3}$$

It can also be referred to as a DC component if

$$f = 0 \qquad (14.4)$$

Here, f_{Fund} represents fundamental power system frequency, f represents frequency, and n > 0 is an integer. Table 14.1 gives a depiction of a list of power quality issues which varies in severity. Based on the level of damage

Table 14.1 Power quality issues.

Power quality issues	Causes	Effects	Severity level
Spinning Reserve	Load Shedding	System operators use this when there a sudden change in load or during outages, to respond to the events	Mild
Flicker	Changes/ fluctuations in voltage supply	The equipments present on the load side are damaged	Moderate
Transient	Lightning, RLC snubber circuits, PE commutation	Disturbances in electrical equipment	Catastrophic
Voltage Spike	Badly regulated transformers, badly dimensioned power sources or stop/start heavy loads	Damage or stoppage of sensitive equipment, flickering of screens and lighting and data loss	Severe
Harmonics	When a sinusoidal voltage is imposed on a non-linear load	Garbled data, lock-up, overheated motors or transformers, loss in electrical equipments	Moderate

(Continued)

Table 14.1 Power quality issues. (*Continued*)

Power quality issues	Causes	Effects	Severity level
Power Frequency Variations	Heavy Load	Major effect on sensitive devices and motors	Mild
Waveform Distortion	Noise in the system	Saturation and overheating of transformers	Moderate
Noise	Improper grounding, electromagnetic interference	Data loss, disturbances in equipement sensitivity	Mild
Long Time Voltage Interruption	Control malfunction, insulation failure or failure to protect the devices	Malfunctioning in the data processing equipments	Moderate
Voltage Swell	Inadequate wiring, In rush current, source voltage variation, stop/start of heavy loads	Garbled data, intermittent lock-up, damage of equipments, data loss	Mild
Voltage Sag	Inadequate wiring, In rush current, source voltage variation, large network loading, fault in the system	Garbled data, intermittent lock-up, overloading issues	Moderate
Voltage Fluctuations	Load Switching	Flickering of lighting, under/over voltages	Severe

and economic cost of it, the ranking of the issues are classified in the table below.

Based on the analysis done in the table, it is observed that transient is said to have a catastrophic impact on the electrical equipment, resulting in damaging them, leading to economic discrepancies.

Source power quality can be defined by two perspectives, namely, internal sources and utility sources. Internal sources include medical equipment, arc welders, lightning ballasts, and individual loads. Similarly, utility sources include switching, faults, power factor correction equipment, and lightning. Some of the other sources are UPS, large motors, electronic dimming system, microgrid, smart grid, battery chargers, and variable frequency drivers. The issues of source power quality are as defined in Figure 14.3.

Figure 14.4 shows a comprehensive outline of the facilities affected by power quality based on the cost considerations.

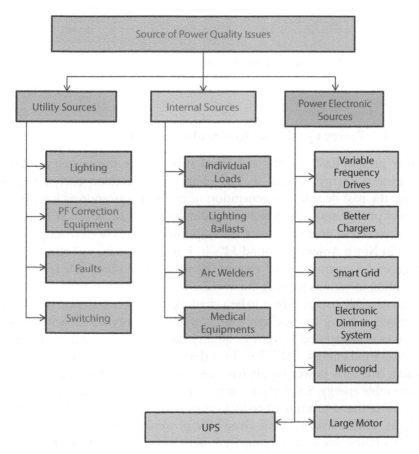

Figure 14.3 Sources of power quality issues.

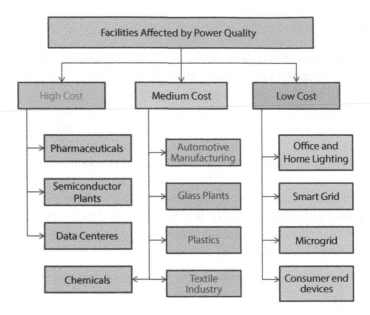

Figure 14.4 Facilities affected by power quality.

14.2.1   Power Quality in Renewable Energy

Solar photovoltaic systems and wind turbines are the primary focus of renewable energy sources in this paper and are clean energy sources. Over the past decade, the generation of green energy sources have been a global pursuit and is also increasing day by day. A survey in 2015 showed that renewable energy sources are responsible for 27% electricity generation in North America, around 34% in Europe and 52% in Latin America. Figure 14.5 portrays a brief view of the variation in renewable energy contribution in various parts of the world.

Renewable energy proves to be a challenge to power systems even though it excels at being an excellent source of clean energy. The discussion that follows elaborates on the power quality issues faced by solar PV panels as well as wind energy. EMF27 in 2017 demonstrated that during the past 50 years, over 35% power supply has been provided by the various sources of renewable energy. Similarly, a study by Oman from Masirah Island, which indicates that a prominent reduction in cost of energy could be observed (75%), when a hybrid Wind-Diesel-PV power system is used. Moreover, if such a renewable energy integrated methodology is used to replace diesel plants, it is possible to slash the greenhouse emission by 25%.

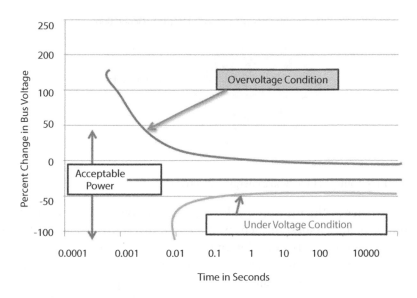

Figure 14.5 Condition of power voltage in computer devices.

By the end of 2011, a total of 4 GW of wind capacity was accumulated by the offshore turbines in Europe. Figure 14.8 represents a typical wind power plant along with a more detailed layout of the connection between the grid and the wind turbine. Commercial, residential, or solar PV plants are used to make a grid-connected solar PV power system. This system will be able to create power that is more than 100 MW. The distributed networks are connected with the commercial and residential PV cells while the transmission systems are connected with the large power plants. In 2009, a total of 20 TWh of power was produced while a total 138.9 GW was produced in 2013. To balance frequency and shedding of load, SR is given in the power system, thereby making it possible to achieve high-quality energy sustainability. In general, there are two types of power quality issues when using renewable energy systems: harmonics and frequency fluctuations and the causes of these power issues are as follows:

- Disturbances in power grid.
- Non-controllable variable renewable energy resources.

In renewable energy generation, the power electronic converters used produce harmonics. This can be identified by comparing with the IEEE Standard 929-2000 which uses four parameters, namely, distortion, frequency, voltage flicker, and voltage to measure the power quality issue. Any

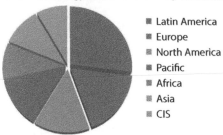

Renewable Energy Sources in Electricity Generation

- Latin America
- Europe
- North America
- Pacific
- Africa
- Asia
- CIS

Figure 14.6 Sources of renewable energy in electricity.

discrepancies in the measured value against the standard value will result in power quality issues. In such instances, an inverter is suggested to prevent this power quality deviation in the utility lines. There are two aspects in addressing this problem, from the power grid side and renewable energy side. Both solar photovoltaic enery and wind energy generation will be affected by intermittency that is caused because of uncertainity features of solar and wind resources. Due to the incident solar energy as well as the change in speed of the wind, the outputs of the renewable energy sources will differ. A representation of the various renewable energy contributed by countries from around the world is shown in Figure 14.6.

14.2.2   Power Quality Issues in Wind and Solar Renewable Energy

14.2.2.1   Wind Renewable Energy

One of the most successful forms of renewable energy is wind energy. The major issues involved in the process of producing power are as follows:

- Voltage Stability: A flexible AC transmission system (FACTS) is used to handle voltage stability issues. Because of surplus or lack of active and reactive power, islanding of connected IG disrupts frequency and voltage. One of the best suggested strategies to overcome this power issue is to regulate the blade pitch to control the turbine.
- Stability Issue: Stability is another aspect that will be affected by energy storage. This is possible by using DFIG generators to replace the SCIG when battery energy storage is added.
- Thermal Loading: Power controllability and reliability are two vital performance parameters in wind power generation. However, as the turbines are built to help fault recover and to

withstand grid faults. However, the low voltage ride through (LVRT) will have a larger thermal stress on the energy generation, thereby affecting the reliability of the system.

14.2.2.2   Solar Renewable Energy

Some common power quality issues in solar renewable energy are as follows:

- Slow and fast voltage variation and voltage imbalance: In the user's side and supplier's side, the remnants vary. Line-to-neutral and line-to-line are identified to be the two voltage points to be measured and any variation in them is compared with the IEC and EN standards to evaluate voltage characteristics and determine its impact on the power generation system.
- Voltage Sag, Swell, and Flicker: Use of static var compensators, large motor starting transients, switching of power factor correction capacitors, and temporary faults are caused generally due to interruptions, surges, flickers, swells, and voltage sags.

14.3   Mitigation of Power Quality Issues

There are two methods to mitigate power quality issues in renewable energy—from the utility side or from the customer side. The approach from the customer side involves installing line conditioning systems which will be able to counteract or suppress the disturbances that occur in power generation. Similarly, at the utility side, it is known as load conditioning where the devices are built such that they are not affected by power disturbances upto a particular level, even in voltage distortion. In order to mitigate PQ problems, a number of devices such as harmonic filters, transient voltage surge suppressors, isolation transformers, noise filters, constant voltage transformers, energy storage systems, super-capacitors, and flywheels. CPDs such as UPQC, DVR, and DSTATCOM are also said to be effective in mitigating PQ power issues.

14.3.1   UPQC

This method is the integration of shunt and series type active filters which are linked together by a common DC capacitor. To mitigate supply side disturbances, UPQC series components are used. Some of the common disturbances include harmonics, voltage unbalance, flicker, voltage swells, and voltage sags. Voltage is inserted in order to protect the load voltage at the required range, distortion free, and balanced. Mitigation of current quality problems are addressed by the shunt component. The issues include load unbalance, load harmonic currents, and poor power factor. This device injects the current such that it can balance with the sine wave and is also in sync with the source voltage.

14.3.2   DVR

A type of series-connected CPD is the DVR. Side disturbances at the supply end are shielded from flowing into sensitive loads. They are also used to isolate the source from the harmonics that are caused by loads, behaving like a series active filter. It is made up of a PWM converter with DC capacitor, passing through a coupling transformer and a low pass filter, in serial connection with a utility supply voltage as shown in Figure 14.7. This device operates such that when there is distortion or unbalance in the source voltage, it will restore the load-side voltage to the required waveform and amplitude by injecting ac voltages that are controlled in synchronism and series with distribution feeder voltages.

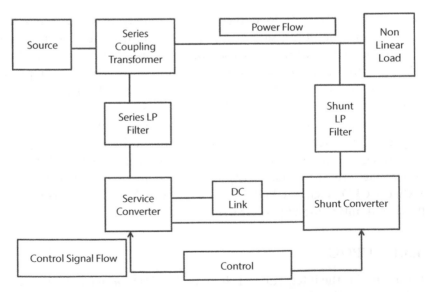

Figure 14.7 Mitigation of power quality issues using UPQC.

14.3.3   D-STATCOM

A D-STATCOM [21] is a CPD which is used for load balancing, current harmonics filtering and power factor correction. This CPD is also known as parallel or shunt active power filter and can also be used in distribution bus to regulate the voltage. The device consists of a pulse wave modulator converter. It is used to inject harmonic components, resulting in 180° phase shift, and also operates as a current controlled PWM converter. When used in the apt configuration, this can also be used to rectify poor load power factor. Figure 14.8 shows a typical D-STATCOM used for mitigating power quality issues.

14.3.4   UPS

Unlimited power supply (UPS) is used to provide a back-up power when there is sudden power failure. The UPS is generally used in places where data loss plays a crucial role, in devices such as computers and telecommunication devices. Double conversion topology is used in good quality UPS where the AC power at the incoming node is converted into DC power to charge the UPS battery. Later on, it can be used to generate AC sine wave that can be used. There are two types of UPS which are commonly used:

- Passive-standby UPS also known as Offline UPS which can be used in applications which require less than 2-kVA power rating.
- Power correction is done with the help of DC UPS.

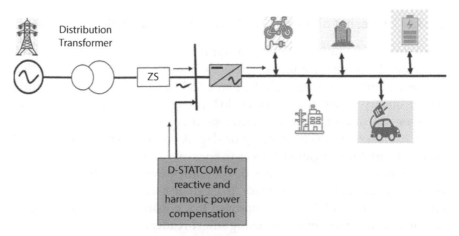

Figure 14.8 Mitigation of power quality issues using D-STATCOM.

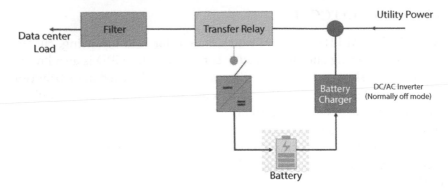

Figure 14.9 Mitigation of power quality issues using UPS.

A UPS is generally built on reduced switch-count configuration wherein seamless transition and active active front-end. Similarly, backward power feed is absent when isolation of grid is ensured at the time of power failure. Thus, a circuit breaker is not necessary when a UPS is used. Figure 14.9 shows the UPS configuration for mitigating power quality issues.

14.3.5   TVSS

To address the serious issues of power quality, TVSS gives an interface between sensitive load and power load. In order to prevent excessive line voltage, a non-linear resistance component like the zener diode is used with TVSS. This will help conduct impulse that is excessive to the ground. In general, TVSS systems are built with shunt protectors at output as well as the input.

14.3.6   Internet of Things in Distributed Generations Systems

The Internet of Things (IoT) plays a significant role in improving the processing and sensing capabilities of devices. Transmission control is handled by the grid interface, earning it the name "self-healing" grid. In mitigating power quality issues, the use of an IoT interfaced grid will be useful to monitor real-time disturbances caused such as power cuts, harmonic interference, flickers, and voltage swing or slag. A control software is designed to link it with the mitigation devices to ensure that in case of serious power interference, the grid can control or route the power such that there is no or minimal disruption taking place. IoT can be used to monitor renewable energy, meet market demand, improve QoS, enhance smart grid services, and also read power consumed using smart readers.

14.4   Discussions

Under-voltage and loss of efficiency are two of the most serious issues that are viewed in examining power quality issues in electrical devices and equipment. Non-linear loads create harmonics issues, switching load will result in long duration voltage interruptions, insulation failure leads to long time voltage interruptions, initiation of heavy loads will create voltage swell/sag, and load switching leads to voltage fluctuations.

- IEEE Standard P1547 gives the characteristics of microgrid and the power system interconnection.
- IEEE Standard P1409 and IEEE Standard 125-1995 address the issues in power quality and provide apt solutions for the same.
- IEEE Standard 1159-1995 characterizes the general power quality issues.
- IEC Standard 61000-4-15 briefly defines the flicker issue.
- IEEE Standard P1547a and IEEE Standard P1564 address voltage sag problem.
- IEC/TS 61000-3-4 standards, IEC 61000-3-2, and IEEE-519 standards provide guidelines to limit harmonic issues.
- To limit the harmonics issues IEC/TS 61000-3-4, IEC 61000-3-2 and IEEE-159 standards are used.

Figure 14.10 shows the effect of power quality issues in cost consideration. It is found that the transient error will lead to severe damage of the device, resulting in high cost impact while swell has a considerably lower impact. Sag, voltage fluctuations, and interruptions are seen to have almost the same cost impact.

Based on the power quality mitigation measures observed in this chapter, it is found that UPQC provides the apt solution for issues like voltage unbalance, flicker, voltage swells/sags, and harmonics when connected in series. However, when connected in shunt, it will be able to protect the device against load unbalance, load harmonic currents, and power factor. On the other hand, current harmonics and poor power factor is mitigated with the help of D-STATCOM. In case of shortage of power, the UPS will play a vital role in applications like telecommunication and electronic devices. TVSS is ideal for correcting voltage transients. Based on the type of power quality issue, the apt device to be used to mitigate the issue is chosen using IoT.

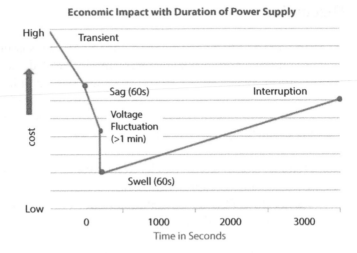

Figure 14.10 Economic impact vs. duration of power supply.

14.5   Conclusion and Future Scope

In recent years, the penetration of distributed generation in the grid has become more prominent. Power quality issues are considered to be a serious threat and end users are finding it to be a more sensitive issue as we move toward perfection. This proposed chapter gives a detailed discussion on power quality issues prevalent in renewable energy sources—wind energy and solar energy. It covers a detailed study on the various sources, the parameters which influence them, and the quality issues faced. A study of CPDs and their impact on power quality issues mitigation is addressed and the role of the IoT in grid distributed generation. An extensive study on the different CPDs is carried out and a comparison chart of the same is also observed. They play a crucial role in ensuring that there are no discrepancies in power quality deliverance at both ends.

There are many future research aspects in power quality issues and scope is very wide. Some of them are listed below:

- The SR level must be designed carefully and it should be examined to ensure that it does not affect the power quality of the system. This includes monitoring of unexpected events like intermittency of renewable sources, sudden load change, and inappropriate SR levels.
- Power quality issues such as instability and loss increase with interaction between utility grids and microgrids. Monitoring

of these issues and appropriate mitigation of the same also need to be analyzed.

- There is a need for more research in inter-harmonics and higher order harmonics mitigation as well as detection. Future work can also include observation of these two issues which contribute a significant amount to power quality.
- Monitoring devices for transients, interruptions, noise, swell, voltage sage, etc., are required to be developed to detect or track the power quality issues in a more detailed manner. Wavelet domain analysis as well as time domain analysis can be performed and the methodologies can be compared to find the more efficient one.
- Monitoring techniques and mitigation of voltage spike have not been dealt with in detail in this book chapter. However, the scope of the same is quite vast and can be discussed further. For the fast varying voltage, proper detection, tracking and monitoring methods need to be introduced.
- A detailed analysis of the hydropower system and the power quality issues involved need to be made. A widespread investigation on mitigation techniques and power quality issues need to be observed and analysed in a more detailed manner.

References

1. Zhao, Y., *Electrical Power Systems Quality*, Academia, University of Buffalo, Nov. 11, 2016, [Online]. Available: http://best.eng.buffalo.edu/Research/Lecture%20Series%202013/Power%20Quality%20Intro.pdf.
2. Yıldız, C., Keçecioğlu, Ö. F., Açikgöz, H., Gani, A., Şekkeli, M., Power quality measurement and evaluation of a wind farm connected to distribution grid. *Procedia-Soc. Behav. Sci.*, 195, 2370–2375, Jul. 2015.
3. Thallam, R.S. and Heydt, G.T., Power acceptability and voltage sag indices in the three phase sense, in: *Proc. IEEE Power Eng. Soc. Summer Meeting*, Jul. 2000, pp. 905–910.
4. Khalid, S. and Dwivedi, B., Power quality issues, problems, standards & their effects in industry with corrective means. *Int. J. Adv. Eng. Technol.*, 1, 2, 1–11, 2011.
5. Krishan, O. and Suhag, S., An updated review of energy storage systems: Classification and applications in distributed generation power systems incorporating renewable energy resources. *Int. J. Energy Res.*, 43, 12, 6171–6210, 2019.

6. Whaite, S., Grainger, B., Kwasinski, A., Power quality in DC power distribution systems and microgrids. *Energies*, 8, 5, 4378–4399, 2015.

7. Ghosh, A. and Ledwich, G., *Power Quality Enhancement Using Custom Power Devices*, Springer, New York, NY, USA, 2012.

8. Bhavani, R., Prabha, N.R., Kanmani, C., Fuzzy controlled UPQC for power quality enhancement in a DFIG based grid connected wind power system, in: *Proc. Int. Conf. Circuit, Power Comput. Technol. (ICCPCT)*, Mar. 2015, pp. 1–7.

9. Yuvaraj, V., Deepa, S.N., Rozario, A.P.R., Kumar, M., Improving grid power quality with FACTS device on integration of wind energy system, in: *Proc. 5th Asia Modeling Symp. (AMS)*, May 2011, pp. 157–162.

10. Ilango, K., Bhargav, A., Trivikram, A., Kavya, P.S., Mounika, G., Nair, M.G., Power quality improvement using STATCOM with renewable energy sources, in: *Proc. IEEE 5th India Int. Conf. Power Electron. (IICPE)*, Dec. 2012, pp. 1–6.

11. Kazmierkowski, M.P., Power quality: problems and mitigation techniques [Book News]. *IEEE Ind. Electron. Mag.*, 9, 2, 62–62, 2015.

12. Liang, X. and Andalib-Bin-Karim, C., Harmonics and mitigation techniques through advanced control in grid-connected renewable energy sources: A review. *IEEE Trans. Ind. Appl.*, 54, 4, 3100–3111, 2018.

13. Nagarajan, A., Sivachandran, P., Suganyadevi, M.V., Muthukumar, P., A study of UPQC: emerging mitigation techniques for the impact of recent power quality issues. *Circuit World*, 47, 1, 11–21(11), 2020.

14. Arya, S.R., Singh, B., Niwas, R., Chandra, A., Al-Haddad, K., Power quality enhancement using DSTATCOM in distributed power generation system. *IEEE Trans. Ind. Appl.*, 526, 5203–5212, 2016.

15. Kathiresh, M. and Subahani, A.M., Smart Meter: A Key Component for Industry 4.0 in Power Sector, in: *Internet of Things for Industry 4.0. EAI/Springer Innovations in Communication and Computing*, G. Kanagachidambaresan, R. Anand, E. Balasubramanian, V. Mahima (Eds.), Springer, Cham, 2020.

16. Ravinder, K. and Bansal, H.O., Investigations on shunt active power filter in a PV-wind-FC based hybrid renewable energy system to improve power quality using hardware-in-the-loop testing platform. *Electr. Power Syst. Res.*, 177, 105957, 2019.

17. Shair, J., Xie, X., Yan, G., Mitigating subsynchronous control interaction in wind power systems: Existing techniques and open challenges. *Renewable Sustain. Energy Rev.*, 108, 330–346, 2019.

18. Bajaj, M. and Singh, A.K., An analytic hierarchy process-based novel approach for benchmarking the power quality performance of grid-integrated renewable energy systems. *Electr. Eng.*, 102, 1153–1173, 2020.

19. Singh, A.K., Kumar, S., Singh, B., Solar PV energy generation system interfaced to three phase grid with improved power quality. *IEEE Trans. Ind. Electron.*, 67, 5, 3798–3808, 2019.

20. Rezkallah, M., Chandra, A., Hamadi, A., Ibrahim, H., Ghandour, M., Power quality in smart grids, in: *Pathways to a Smarter Power System*, pp. 225–245, Academic Press, London, UK, 2019.
21. Rohouma, W., Balog, R.S., Peerzada, A.A., Begovic, M.M., D-STATCOM for harmonic mitigation in low voltage distribution network with high penetration of nonlinear loads. *Renewable Energy*, 145, 1449–1464, 2020.

15

Smart Grid Implementations and Feasibilities

Suresh N. S.[1]*, **Padmavathy N. S.[2]**, **S. Arul Daniel[1]**
and Ramakrishna Kappagantu[3]

[1]Electrical and Electronics Engineering Department, St Joseph Engineering College, Mangalore, India
[2]Civil Engineering Department, National Institute of Technology, Tiruchirappalli, India
[3]POWER GRID. Smart Grid Pilot Project, Pondicherry IEEE R10 Nominations & Advisory (2017-2018), IEEE R10 (Asia Pacific) (2015-2016) Bangalore, India

Abstract

Electric power has become the backbone of the economic growth of any nation, across the globe. In the world's total electrical energy production 66% was from coal, gas, and oil and this power infrastructure is bulky, complex, and non–eco-friendly. Global smart grid federation (GSFG) explained how the existing power grid network is not capable to meet the requirements of the 21st century. At present, the power network structure is following the concept of centralized power generation. Smart grid (SG), the new facet of power infrastructure, facilitates the decentralization of power generation. An SG has shown its competence to meet power availability, reliability, quality, economic operation, efficiency, safety, security, and other essential parameters along with environmental issues. The concept of SG, its technologies, challenges, and difficulties in implementing have been discussed in detail on several platforms. It is better to understand the techno-economic issues of SG infrastructure for future advancement and the expansion of SG projects. For effective planning, it is essential to understand both the positive and negative aspects of the SG technologies. Implementation of any technology reflects in terms of investment sought and its success is measured by profit gained by both investor and the society.

*Corresponding author: drsureshns@gmail.com

M. Kathiresh A. Mahaboob Subahani and G.R. Kanagachidambaresan (eds.) Integration of Renewable Energy Sources with Smart Grid, (327–346) © 2021 Scrivener Publishing LLC

Keywords: Smart grid, pilot project, smart meters, implementation, power quality, solar PV

15.1   Introduction

A smart grid (SG) has certainly emerged as a logical solution to all the issues of the power industry. SG has shown the great potential of smart management of resources, integration of renewables, efficient and transparent market operation, the involvement of consumers in grid management, etc. SG is redefining the power industry by transforming the centralized power network into a decentralized one with a specific focus on generation and distribution. Efforts for deploying SG have witnessed momentum on a global scale. India too has opted for SG as a solution to resolve power deficit and other management issues. As its first significant initiative, India has launched 14 SG pilot projects across the country [8].

15.1.1   Smart Grid Technologies—Literature Review

A study of the literature reveals that instead of being a technological term, the SG gains more significance as a marketing term that encapsulates various technologies. El-Hawary [1] gives a concise definition of smart-grid as "the smart grid is a suit of an information-based application made possible by increased automation of the electricity grid, as well as the underlying automation itself; this suite of technologies integrates the behavior and actions of all connected supplies and loads through dispersed communication capabilities to deliver sustainable, economic and secure power supplies". IRENA [13] discusses the four basic functions of the SG as a resource, controller, collector, and assessor/display, and for this purpose, SG uses various technologies. The list of functionalities that SG incorporates in the power network includes the following:

a.   Advanced metering infrastructure (AMI)
b.   Distribution automation (DA) and distribution management system (DMS)
c.   Distributed storage
d.   Microgrid
e.   Virtual power plant (VPP)

    f.   Smart inverter
    g.   Information and communication technologies (ICTs)
    h.   Cybersecurity
    i.   Geographical information system (GIS), etc.

Besides numerous advantages, major applications of SG technologies are as follows:

    a.   Advance electricity pricing
    b.   Demand response (DR)
    c.   Outage management system (OMS)
    d.   Data collection, management, and utilization
    e.   Energy accounting
    f.   Energy security
    g.   Green power, etc.

Literature also reveals how the SG and renewables are enabling each other. A SG makes way for the integration of renewable sources in the power network and renewables sources support the grid network in terms of green energy. Further, this work discusses, in brief, various global as well as Indian initiatives of various stakeholders for the deployment of the SG.

15.2 Need for Smart Grid

The electricity sector is confronted by critical challenges, *viz.*, growing energy demand, high AT&C losses, concern on reliability and quality of power supply, fuel constraints, and implementation of environmental policies to combat climate change, etc. The open electricity market has given a new paradigm shift to the way power is generated, transmitted, and distributed. These challenges are leading to increased development of electric power production from grid-interactive renewable energy sources which are inherently volatile, exploration of new possibilities for energy storage, development of electric vehicles (EVs), recognition of consumers and utility as smart energy decision-makers, and advancement of energy efficiency in real time, etc. [9].

Realizing increased complexities in every segment of the electricity value chain, it is prudent to introduce intelligence in all fronts of the power sector through SG applications which may facilitate efficient and reliable end-to-end intelligent two-way delivery system from source (Generation) to sink (Consumer Load) through the integration of renewable energy sources, smart transmission, and distribution. In this way, SG technology shall bring efficiency and sustainability in meeting the growing electricity demand with reliability and the best of quality. It would also empower the consumer to participate in the energy management process. The acceptance and transformation will require significant effort, investment, and allocation of resources to demonstrate a new "Smart Grid" infrastructure for integration of renewable energy sources as well as the adoption of smart tools to manage electricity flows, energy efficiency, conservation, and use [9].

15.2.1   Smart Grid Description

As depicted in Figure 15.1, SG is an interconnection of modern communication and information technologies along with the control and automation process across the entire electricity sector comprising of generation, transmission, distribution, and the consumer. SG aims to make existing infrastructure more robust, reliable, and efficient by using intelligent tools and technologies, encourage active demand-side participation, ensure the sustainability of supply through renewable integration, promote green and clean environment, and provide incentives for efficient production, transmission, distribution, and consumption of electricity across the value chain of electricity.

Toward generation, new technology and high capacity plants are added every year which poses operation challenges. For sustainability, huge focus is on renewable integration which demands new and modern

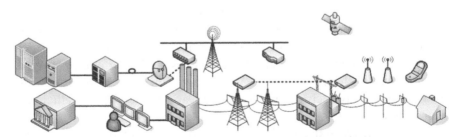

Figure 15.1 Interconnection of power, IT, and communication technologies in smart grid.

technology (like storage and demand response) for its integration with the grid.

In transmission ultra-high voltage, 1,200-kV AC and 800-kV HVDC networks are being built for long-distance bulk power transfer requiring efficient use of power electronics devices (STATCOM/SVCs, etc.). Deregulation has encouraged wholesale power transactions providing opportunities for electricity generators to transfer electricity outside of the original service areas to respond to market needs and opportunities. This has stressed the transmission system beyond the limits for which it was designed and built. Managing these complex networks requires real-time monitoring and control with state of art technologies (SCADA, Synchrophasor, etc.) along with robust information and communication networks.

In distribution, comprehensive changes are happening in the power sector. Distribution utilities are going through a reformation process for the reduction in AT&C loss, energy conservation, improvement in customer satisfaction, power quality, reliability of power supply, etc. Rationalization of the tariff, based on time-of-use, real-time pricing, etc., is taking shape. Achieving all these functionalities in distributions requires the latest metering and measurement technologies (like AMI, OMS, and PQM).

Therefore, SG is a confluence of information, communications, and electrical/digital technologies, facilitating a seamless integration of business processes and systems to yield real measurable value across the entire power delivery chain.

15.3   Smart Grid Sensing, Measurement, Control, and Automation Technologies

Monitoring and control in the power sector have evolved with the installation of Remote Terminal Unit (RTU), SCADA systems along with communication facilities to monitor and control the grid station of transmission/distribution network. Transducers and advance digital signal processing technologies were used to measure and monitor the electrical parameters at control centers. These capabilities are now extended to the distribution area through demand-side management to optimize demand, shifting of peak energy, energy, and network availability. AMI is installed to monitor, manage, and control the load at consumer premises.

15.3.1   Advanced Metering Infrastructure

Electricity meters are used to measure the quantity of electricity supplied to customers and calculate energy and transportation charges for electricity retailers and network operators. An accumulation meter is one that records the consumption of energy over time. In recent years, industrial and commercial consumers with large loads have increasingly been using more advanced meters, for example, interval meters (special energy meters used for ABT) which record high energy over short intervals, typically every 15 minutes.

After the accumulation meter, another generation of meter commonly known as AMR (Automatic Meter Reading) meters was introduced, where energy consumption information is being transferred periodically (say monthly) from the meters to the energy supplier and/or network operator using low-speed one-way communication networks. AMR has given way to more advanced two-way communication infrastructure AMI, which supports applications like varying tariffs, demand-side management, remote connect/disconnect, online energy consumption information, etc.

15.3.2   Key Components of AMI

AMI refers to systems that measure, collect, analyze, transmit, and manage energy using advanced ICTs. AMI consists of a smart meter, a bi-way communication circuit, Home Area Network (HAN), Data Concentrator Unit (DCU), Meter Data Management System (MDM), and Utility Data Center.

15.3.3   Smart Meter

Smart meters measure electrical parameters, with high accuracy. Data are stored by the smart meter at specific. It stores data at particular intervals and transmits it to the demand side management center for utility control and other demand management applications. Smart meter is equipped with a local display through which consumer can monitor their consumption pattern. Presently, in general, data transfer from the meter is on the proprietary protocol.

As a standard for data transfer between meters to the DCU/Smart Grid Center, the DLMS/COSEM data transfer protocol is developed. Figure 15.2 shows a typical smart meters. A keypad in the smart meter enables a machine-human interface, through which a user can access the settings of the meter. For these latest meters nowadays have a touch screen

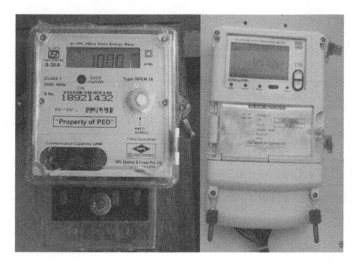

Figure 15.2 Smart meters.

and a user can either control the smart device else he can choose to know the bill details of that smart appliance.

15.3.4   Communication Infrastructure and Protocols for AMI

The different components in AMI are linked together with the communication path and sensor nodes to provide interoperability between them. For AMI information flow, two types of information infrastructure are needed. First, from sensors and electrical equipment to HAN smart meters, and second, from smart meters to utility data centers. The first data flow is done by power line communication or wireless communication, such as ZigBee, RFF, and others. For the second flow cellular technologies like GSM/GPRS/CDMA and the Internet is used [15]. The technology choice depends upon the type of environment.

Typical communications architecture for smart metering comprises of Wide Area Network (WAN) and HAN. A WAN connects the AMI system with the utility network. A HAN is a smart network at the home level. A HAN enables centralized energy management which includes energy monitoring and display, controlling smart appliances and smart plugs, scheduling and remote operation of house-hold appliances, and household security systems for the comfort and convenience of consumers. It is expected that HAN will be beneficial to the utilities in implementing demand response. It would also be helpful in the management of dispersed generation and the charging of EVs. Summary of AMI communication is tabulated in Table 15.1.

Table 15.1 Communication technologies for AMI.

Sl. no.	Communication link	Technologies used
1	IHD to meter	RF
2	Smart Meter to DCU*	RF, PLC, GPRS
3	DCU to MDAS	GPRS, Optical Fiber Networks
4	MDAS to MDM	LAN
5	Consumer access	Web services

*Smart meter based on GPRS can communicate to MDAS directly.

15.3.4.1 Data Concentrator Unit

A data concentrator gathers the data from the smart meters. The data from the smart meters are collected by a DCU as shown in Figure 15.3 and they were then forwarded to the control center. It also sends data for a particular/ all obtained from the utility/customer smart meter to the intended consumer.

MDAS is a critical component of the AMI system. It collects meter data directly from the meter or through DCU and sends load to enable or curtailment signals, connect or disconnect, billing, and other control information to the smart meter. MDAS exchange meter data to meter data management systems coupled with analytics on the standard data exchange

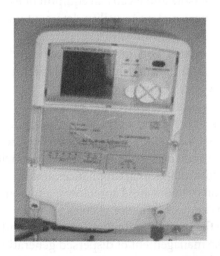

Figure 15.3 DCU.

model. Data Center/Smart Grid Control Center is equipped with hardware to host MDAS and MDMS application. Utility manages the demand side operations through this control center. Figure 15.4 shows interfaces of MDM with different devices/applications.

AMI is the backbone of an intelligent distribution system. Typical distribution management architecture with AMI interface is shown in Figure 15.5.

15.3.5   Benefits of AMI

AMI provides the following benefits:

a.   Availability of energy usage information of each consumer to utility as well as to consumer himself.
b.   Energy audit for a set time interval to check theft and pilferages.
c.   Load/device control from remote is possible by utility and consumer.
d.   Help in outage management up to consumer level.
e.   Consumer engagement in energy management.
f.   Improved billing efficiency.
g.   Enabling Time of Use (ToU) Tariff.

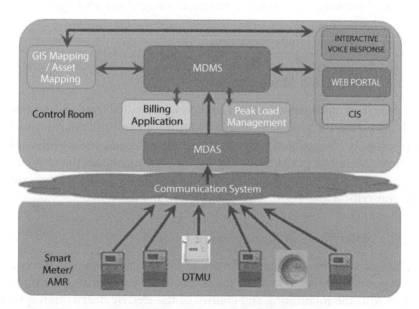

Figure 15.4 Interface of MDM with different devices/applications.

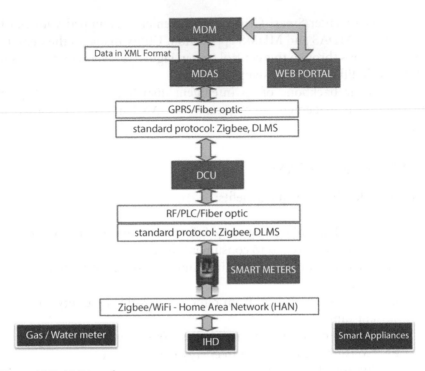

Figure 15.5 AMI interface.

    h. Enabling distributed generation integration through net metering.

    i. Real-time tamper detection.

15.3.6   Peak Load Management

The peak load management (PLM) collects information from different units such as load forecasts, SCADA, and MDM subsystems. With this information power supply is scheduled for present and future needs, meanwhile, deficit or surplus details are calculated. The device then aggregates the response of the consumers to these signals. PLM will also involve load shifting, critical and non-critical load segregation, etc., as seen in Figure 15.6.

15.3.7   Distribution Management System

A DMS monitors and controls the whole distribution network. It acts as a decision support system to assist the control room and field operating personnel with the monitoring and control of the electric distribution system. DMS access real-time data and provide all information on a single console at

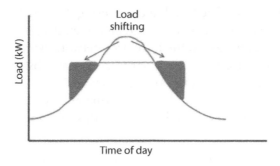

Figure 15.6 Peak load management.

the control center in an integrated manner. DMS grew initially from SCADA systems but it performs more functions with better visualization and intelligence than the SCADA system [3]. DMS will result in the following:

a. Reduction in the duration of outages.
b. Improvement in speed and accuracy of outage predictions.
c. Reduction in crew patrol and drive times through improved outage locating.
d. Determine the crew resources necessary to achieve restoration objectives.
e. Effectively utilize resources between operating regions.
f. Determine when best to schedule mutual aid crews.
g. Increased customer satisfaction.
h. Provide customers with more accurate estimated restoration times.
i. Improve service reliability by tracking all customers affected by an outage, determining electrical configurations of every device on every feeder, and compiling details about each restoration process.

Thus, the DMS is more advanced than the SCADA system and results in a decrease in outage time, outage cost, and reliable power supply to consumers.

15.3.8   Distribution Automation System

DA systems include automation of substation and automation of feeder. Automation of substation includes supervisory control and monitoring of circuit breakers, load tap changers (LTCs), regulators, and substation capacitor banks. Remote data acquisition is required to achieve effective use of the supervisor control function. Substation automation system comprises of decentralized architecture using IEC 61850 protocol [2]. Sensors/relays/bay controllers are intelligent electronic devices (IEDs), which shall be connected to the local area network within the substation and with the greater grid system at large. A typical substation automation network within a substation is shown in Figure 15.7.

Automation of feeder includes monitoring and supervisory control of auto-recloser, sectionalizes, and switches. This also includes the remote

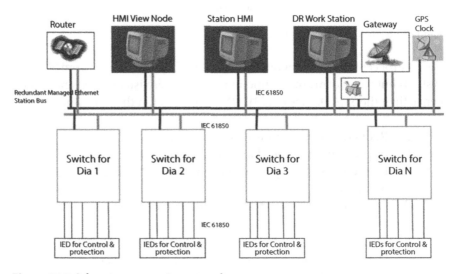

Figure 15.7 Substation automation network.

Table 15.2 Example of typical house hold saving.

Average Power Saving through Energy Efficiency per Household		
	Summer	**Winter**
Peak Demand Reduction (watt)	1,217	2,485
Reduction in Monthly Consumption (kWh)	266	199
Value of Saving per Household		
Average Tariff	RS. 4.00/Unit	
Monthly Saving	Rs. 1,063	Rs. 796
Annual Saving (winter 4 months and summer 8 months)	**Rs. 11,688**	
Annual Air pollution reduction (tonnes of CO2)	**2.60**	

operation of equipment installed at the feeder to reduce outage time depending upon the fault location and feeder configuration without any involvement of humans for opening and closing command.

Automated distribution system is much more than a sophisticated SCADA system; it is a true automation system where software capability will replace the operator in some utility operations, thereby freeing them for other assignments and improving their productivity. This automation has resulted in the considerable energy saving for the consumers. The Table 15.2 shows the monthly average power saving in terms of power and tariff for a single home in an automated distribution system.

15.4   Implementation of Smart Grid Project

Intelligent grid, grid-wise, EPRI's intelligrid, etc., are different names of SG [1], which itself is a marketing term [IEC]. Today, SG has many unified universal definitions as smartness signifies different aspects related to each component of the power system. SG technologies include AMI, DA, DMS, ICT, and microgrid.

A vast literature is available on each of these SG attributes. These technologies have the potential to ensure advanced electricity pricing [3], energy security [20], energy accounting California energy commission [12], clean energy [17], and many such advantages. However, for the advancement of these technologies, there are multiple barriers encountered for the implementation of a SG. Wenye Wang and Zhou Lu [4] have identified the objectives and concerns of cybersecurity in the SG. Pathirikkat Gopakumar et al. [14] explained various stability concerns of the SG. Mohamed E. El Hawary gave a brief summary of economic benefits and new research directions on SG implementation. Srikanth Chandra Sekaran [5] and Arup Sinha et al. [6] surveyed and explained about challenges in SG deployment in India.

15.4.1   Challenges and Issues of SG Implementation

Despite the remarkable capabilities that SG possesses to resolve several complex issues in the power industry, its implementation is not an easy task. Implementation barriers are not only limited to technical challenges associated with emerging technologies but also extend to socio-economic factors as well as issues that arise due to the lack of policies and awareness. The objective of this work is to identify these specific barriers and briefly analyze their impact on SG attributes and deployment. Implementation barriers are broadly classified into three categories: technical challenges, socio-economic challenges, and miscellaneous [10].

Technical challenges are like inadequate grid infrastructure that becomes a roadblock to SG deployment in developing countries like India [15]. Integration of SG functionalities with ICTs, *viz.*, big data, 5G, internet, and protocols, certainly, enhances the capabilities of the power network, market, and management, but simultaneously, it may create great concerns on cybersecurity [4]. Storage technologies are essential for seamless integration of renewables, but their respective capabilities, efficiencies, and costs pose serious issues for SG developers. Large-scale deployment of AMI and other technologies has resulted in huge data handling issues, *viz.*, data storage, analysis, and management. The communication industry has developed multiple solutions for communication issues, but still, a common universal, cost-effective solution appears to be out of reach. Adoption of renewables and storage has enough potential to solve the issues of green power and power deficiency, but their large-scale integration into the grid raises a serious threat to grid stability and economics, if not properly addressed. Deploying EV to meet peak load demand is an excellent proposal, but the energy management of these EVs could develop as a major technical obstacle.

High capital investment is one of the major social-economic challenges faced by SG deployment. Similarly, stakeholders play a major role in the success of SG implementation, but factors like new technology, high capital investment, lack of accurate information, and awareness may create a negative perception among these stakeholders. System operational aspects like manpower utilization, billing, tariff structure, and operational strategies can be improvised through SG deployment by proper policy and grid code development. Awareness programs among the consumers to properly understand the benefits of SG are of utmost importance because myths and lack of awareness may also create a negative impact on SG implementation. SG handles a huge quantum of consumer data, and in the absence of well-defined guidelines, policies, and regulations on cybersecurity, it may cause major concern about consumer privacy. A proper understanding of new communication and IT development, *viz.*, 4G, 5G, advanced database management and storage systems, and cloud initiatives, may naturally drive away from the fear cloud of obsolesces among the consumers. Despite several benefits offered by SG, lack of awareness induces fear in consumers that the integration of SG technology would result in increased charges of their electricity bills. Observations on concerns of few consumers and medical groups about the impact of radio frequency (RF) signal deployed in SG, on humans, and health are also underscan and studies of research by some stakeholders may certainly clear such doubts and pave the clearer way to implement successful smarter distribution network.

Miscellaneous issues include regulations and policies. The SG is a relatively new technology, and the development of suitable policies and regulations for smooth and successful implementation of SG are still in the pipeline. Apart from technical and policy level initiatives, since developing countries like India are facing serious issues of power theft and mismanagement, social initiatives are also essential to change the mindset of concerned stakeholders involved. Inadequate experienced workforce and coordination among stakeholders could also become a bottleneck, which can be overcome through skill development programs for successful learning and adoption of SG implementation.

15.4.2 Smart Grid Implementation in India: Puducherry Pilot Project

Puducherry SG pilot project is one of the 14 proposed Indian pilot projects and developed jointly by POWERGRID and PED [8]. The motive behind developing these pilot projects is to evaluate real benefits and identify suitable technologies/models of SG for the Indian perspective [13]. Puducherry SG pilot project area in division-1 of Puducherry has 100% electrification and includes 33,389 nos. of consumers with about 59% of domestic consumers and rest being a combination of commercial, HT, agriculture, street, lighting, etc. [11].

15.4.3 Power Quality of the Smart Grid

The SG comprises and contributes smartness to the power system network required for integrating conventional power sources as well as nonconventional and renewables sources like solar PV, wind, and storage devices. Apart from industrial and domestic consumers, the SG also caters to new and emerging loads like EVs, smart homes, and numerous power electronic loads. All these load components exhibit different characteristics of operation and may result in poor power quality which otherwise mandates proper monitoring and control. Besides serving diversified sources and loads, the SG hosts multiple interfacing devices also which may inject harmonics in the power system network. The increased population of power electronic components means increased harmonic distortion. Jukka V. Paatero et al. [16] and Joakim Widen et al. [7] describe how the PV integration into power system network influences voltage as well as network losses. Monitoring and analysis of power quality enable the system operator to identify and eliminate grid disturbances before they culminate into any major interruption.

15.5   Solar PV System Implementation Barriers

With a 1.27 billion population and an economic growth rate of 7.4%, India is on a significant growth trajectory. The dual requirement of non-fossil energy and the sustainable environment has put the focus on renewables like wind and solar. Further, the urgency and opportunity factors got added to give the biggest ever push for Indian policymakers to "go green" and enhance the contribution of renewables. Central and state authorities are encouraging the modification of many SG pilots and rooftop solar power generation on the demand side projects through subsidies and incentives. National action plan for climate change (NAPCC), Jawaharlal Nehru national solar mission (JNNSM), Indian SG vision and roadmap, etc., are some major initiatives taken by the Indian government in this regard.

Vision and road map of SGs, announced by the Ministry of Power (MoP) in 2013 [18], and policies mandating rooftop solar PV installations were envisaged for broad footprint facilities, i.e., those with a connected load greater than 20 kW or otherwise specified threshold [18]. India's Ministry of New and Renewable Energy (MNRE) has taken good effort to accelerate the growth of 60 towns/cities (designated Solar Cities) [19]. Puducherry also comes under the purview of the MNRE's ambitious solar city program [11].

India is projected to invest $44.9bn in smart metering, DA, battery storage, and other SG market segments over the next decade. This investment will help to reduce the country's staggering 22.7% transmission and distribution loss rate. It has the second largest electricity customer market size in the world. Unlike China, which has the largest, the Indian market will be open to international vendors, as stated in the central government's SG development strategy. This will create very significant market opportunities for leading global players. Vendors from across Europe, North America, and Asia have already participated in small-scale pilots and grid upgrade projects and have been linked with announcements of large-scale rollouts by Indian utilities that are upcoming in the next several years.

India has a power sector market conditions that will require significant SG infrastructure investment. It has one of the highest transmissions and distribution (T&D) loss rates in the world. In some states, the T&D loss rates exceed 50%, and almost all states have loss rates above 15%. Most Indian utilities fail to achieve cost recovery, and SG investment will be an important tool for utilities to reduce losses and improve revenue collection and operational efficiency [10].

The Indian central government has taken several measures to support SG development, including financial revitalization programs for utilities, the establishment of a central SG agency, and the publication of recommended financing strategies for early deployments. With strong drivers and a willing government, the medium- to long-term SG opportunities in India are enormous. In the near term, the environment may be more challenging. India's power sector is fragmented and complicated. As in the United States, each state has a regulatory commission. Industry structure and regulations can vary widely in India from state to state. Understanding the dynamics of each state will be critical to participating in this market.

15.6 Smart Grid and Microgrid in Other Areas

Some of the SG and microgrid setups found in universities explores more varied application areas. The universities are proposing a new level of microgrids and SGs. The universities are finding application of these grids in ships, spacecrafts, and server operations [21]. The laboratories mainly focus on new areas and improvising the existing system through SG concepts. Maritime onboard grid and spacecraft application are evolving a lot today due to promising researches. IoT microgrid labs research on future SGs to facilitate cheap and comfortable smarthomes. The models and techniques developed by these facilities harness research areas such as embedded systems, fast computing systems, and IoT systems. These lab facilities can be used as pilot facilities for individuals to get their systems tested, validated, and developed.

15.6.1 Maritime Power System

The power system in a ship is a microgrid. This microgrid network can generate, supply, and store energy to meet the ships load demands. This ship, ferries, or vessels are becoming smart as navigation in sea today depends more on electrical and electronic gadgets. Ships today are more and more electrical and can be considered as a mini SG. Research labs and pilot facilities are concentrating more on developing ship microgrid and hybrid systems to cater the need of modern day maritime vessels [21]. Moreover, the ports anchoring the electric ships are becoming more and more electrical to support electrical needs.

15.6.2 Space Electrical Grids

Satellite systems need high quality and reliable power system networks. SGs are the only effective solution for satellite applications. Electrical

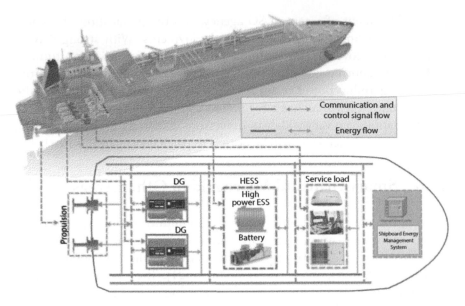

Figure 15.8 Ship board electrical grid.

power systems for satellite-based SGs are reviewed and researched world wide in domains like generation, storage, protection, and reliability [21]. Development of artificial ecosystem for human space exploration is the present important consideration in space missions. Long-term manned missions will not have any interaction with earth. Everything should be self sustained and more importantly the food generation and waste treatment. To maintain a safe artificial ecosystem for astronauts require high-precision grid system; thus, a well-designed SG system will serve the purpose (Figure 15.8).

15.7   Conclusion

Overall, India is one of the most unique SG markets in the world. It combines enormous market potential and a high GDP growth rate with complex regulatory structures and low per capita income. Challenges are certain to persist, but the government's commitment to addressing India's significant power sector challenges by investing in SG infrastructure ensures strong market growth throughout the coming decade. A SG facility in the existing distribution system of Puducherry has been successfully deployed [13]. Preliminary studies and results show improved efficiency and reliability on the distribution side. Distributed computing and communications which deliver real-time information and enable the near-instantaneous balance of supply and demand at the device level have been incorporated. Giving training and making consumers understand better on SG features, *viz.*, load restriction and data consumption display, the consumer can monitor and adjust loads besides saving the energy through demand response. The utility also benefits, *viz.*, demand-side management according to the ToU tariffs, optimal outage management, and integration of renewables by net metering.

References

1. El-Hawary, M.E., The Smart Grid-State-of-the-art and Future trends. *Electr. Power Compon. Syst.*, Taylor & Francis, online, 42, 3–4, 239–250, 5th February 2014.
2. IEC, *Smart Grid: Optimal electricity delivery: What is Smart Grid?*, available at: http://www.iec.ch/smartgrid/background/explained.htm.
3. Hart, D.G., How Advance Metering can Contribute to Distribution Automation, in: *IEEE SMARTGRID Newsletter*, IEEE, online, August 2012.
4. Wang, W. and Lu, Z., Cyber security in Smart Grid: Survey and Challenges. *J. Comput. Networks*, 57, 5, 1344–1371, April 2013.
5. Sekaran, S.C., IEEE-SA pinpoints four key challenges for smart grid implementation in India, in: *IEEE Smart interaction*, 24th October, 2012.
6. Sinha, A., Neogi, S., Lahiri, R.N., Chowdhury, S., Smart Grid Initiative for Power Distribution Utility in India, in: *IEEE general meeting of Power and Energy Society*, July 2011, pp. 1–8.
7. Widen, J., Wackelgard, E., Paatero, J., Lund, P., Impact of distributed photovoltaics on network voltage: Stochastics simulation of three Swedish low-voltage distribution grid. *Electr. Power Syst. Res.*, 80, 12, December-2010.
8. https://indiasmartgrid.org/pilot.php webpage-India
9. https://nsgm.gov.in/en
10. Kappagantu, R., Arul Daniel, S., Suresh, N.S., Techno-economic analysis of Smart Grid pilot project- Puducherry. *Resour.-Effic. Technol.*, 2, 185–198, 2016.
11. Report on "Rooftop solar power systems at Puducherry", Power Grid Corporation of India Limited. (PGCIL), Science Direct- TOMSK Polytechnic University.
12. *Energy Accounting: A Key Tool in Managing Energy Costs; Energy Efficiency Project Management handbook*, Second Edition, California Energy Commission, January 2000.
13. IRENA, *Smart Grid and Renewables: A Guide for Effective Deployment*, November 2013.
14. Gopakumar, P., Jaya bharata Reddy, M., k. Mhanta, D., Letter to the Editor: Stability Concerns in Smart Grid with Emerging Renewable Energy

Technologies. *Electr. Power Compon. Syst.*, 42, 3–4, 418–425, 5th February 2014.

15. Kathiresh, M. and Subahani, A.M., Smart Meter: A Key Component for Industry 4.0 in Power Sector, in: *Internet of Things for Industry 4.0. EAI/Springer Innovations in Communication and Computing*, Kanagachidambaresan, G., Anand, R., Balasubramanian, E., Mahima, V. (Eds.), 2020.
16. Paatero, J.V. and Lund, P.D., Effect of large-scale photovoltaic power integration of electricity distribution networks. *Renewable Energy*, 32, 2, February-2007.
17. https://www.irena.org/
18. https://www.powermin.nic.in/
19. https://mnre.gov.in/
20. https://www.iea.org/
21. Center for Research on Microgrids (aau.dk).

Index